Green Public Procurement Strategies for Environmental Sustainability

Rajesh Kumar Shakya
The World Bank, USA

A volume in the Advances in Environmental Engineering and Green Technologies (AEEGT) Book Series

IGI Global
DISSEMINATOR OF KNOWLEDGE

Published in the United States of America by
IGI Global
Engineering Science Reference (an imprint of IGI Global)
701 E. Chocolate Avenue
Hershey PA, USA 17033
Tel: 717-533-8845
Fax: 717-533-8661
E-mail: cust@igi-global.com
Web site: http://www.igi-global.com

Library of Congress Cataloging-in-Publication Data

Names: Shakya, Rajesh Kumar, 1966- editor.
Title: Green public procurement strategies for environmental sustainability / Rajesh Kumar Shakya, editor.
Description: Hershey, PA : Engineering Science Reference, [2019]
Identifiers: LCCN 2018014733| ISBN 9781522570837 (hardcover) | ISBN 9781522570844 (ebook)
Subjects: LCSH: Government purchasing--Environmental aspects. | Environmentalism.
Classification: LCC JF1525.P85 G739 2019 | DDC 352.5/3--dc23 LC record available at https://lccn.loc.gov/2018014733

This book is published in the IGI Global book series Advances in Environmental Engineering and Green Technologies (AEEGT) (ISSN: 2326-9162; eISSN: 2326-9170)

British Cataloguing in Publication Data
A Cataloguing in Publication record for this book is available from the British Library.

Advances in Environmental Engineering and Green Technologies (AEEGT) Book Series

ISSN:2326-9162
EISSN:2326-9170

Editor-in-Chief: Sang-Bing Tsai, University of Electronic Science and Technology of China Zhongshan Institute, China & Ming-Lang Tseng, Lunghwa University of Science and Technology, Taiwan & Yuchi Wang, University of Electronic Science and Technology of China Zhongshan Institute, China

MISSION

Growing awareness and an increased focus on environmental issues such as climate change, energy use, and loss of non-renewable resources have brought about a greater need for research that provides potential solutions to these problems. Research in environmental science and engineering continues to play a vital role in uncovering new opportunities for a "green" future.

The **Advances in Environmental Engineering and Green Technologies (AEEGT)** book series is a mouthpiece for research in all aspects of environmental science, earth science, and green initiatives. This series supports the ongoing research in this field through publishing books that discuss topics within environmental engineering or that deal with the interdisciplinary field of green technologies.

COVERAGE

- Biofilters and Biofiltration
- Sustainable Communities
- Air Quality
- Contaminated Site Remediation
- Waste Management
- Green Transportation
- Water Supply and Treatment
- Renewable energy
- Cleantech
- Electric Vehicles

IGI Global is currently accepting manuscripts for publication within this series. To submit a proposal for a volume in this series, please contact our Acquisition Editors at Acquisitions@igi-global.com or visit: http://www.igi-global.com/publish/.

The Advances in Environmental Engineering and Green Technologies (AEEGT) Book Series (ISSN 2326-9162) is published by IGI Global, 701 E. Chocolate Avenue, Hershey, PA 17033-1240, USA, www.igi-global.com. This series is composed of titles available for purchase individually; each title is edited to be contextually exclusive from any other title within the series. For pricing and ordering information please visit http://www.igi-global.com/book-series/advances-environmental-engineering-green-technologies/73679. Postmaster: Send all address changes to above address.

Titles in this Series

For a list of additional titles in this series, please visit:
https://www.igi-global.com/book-series/advances-environmental-engineering-green-technologies/73679

Handbook of Research on the Adverse Effects of Pesticide Pollution in Aquatic Ecosystems
Khursheed Ahmad Wani (ITM University Gwalior, India) and Mamta (Jiwaji University, India)
Engineering Science Reference • ©2019 • 538pp • H/C (ISBN: 9781522561118) • US $295.00

Harnessing Marine Macroalgae for Industrial Purposes in an Australian Context Emerging ...
Göran Roos (Economic Development Board of South Australia, Australia) Anthony Cheshire (Balance Carbon Pty Ltd, Australia) Sasi Nayar (South Australian Aquatic Sciences Centre, Australia) Steven M. Clarke (South Australian Research and Development Institute, Australia) and Wei Zhang (Flinders University, Australia)
Engineering Science Reference • ©2019 • 291pp • H/C (ISBN: 9781522555773) • US $195.00

Advanced Methodologies and Technologies in Engineering and Environmental Science
Mehdi Khosrow-Pour, D.B.A. (Information Resources Management Association, USA)
Engineering Science Reference • ©2019 • 506pp • H/C (ISBN: 9781522573593) • US $295.00

Transitioning Island Nations Into Sustainable Energy Hubs Emerging Research and ...
Catalina Spataru (UCL Energy Institute, UK)
Engineering Science Reference • ©2019 • 172pp • H/C (ISBN: 9781522560029) • US $135.00

Recent Challenges and Advances in Geotechnical Earthquake Engineering
T.G. Sitharam (Indian Institute of Science, India) and J. S. Vinod (University of Wollongong, Australia)
Engineering Science Reference • ©2019 • 315pp • H/C (ISBN: 9781522569480) • US $195.00

Agrifood Economics and Sustainable Development in Contemporary Society
Gabriel Popescu (Bucharest Academy of Economic Studies, Romania)
Engineering Science Reference • ©2019 • 401pp • H/C (ISBN: 9781522557395) • US $195.00

For an entire list of titles in this series, please visit:
https://www.igi-global.com/book-series/advances-environmental-engineering-green-technologies/73679

IGI Global
DISSEMINATOR OF KNOWLEDGE

701 East Chocolate Avenue, Hershey, PA 17033, USA
Tel: 717-533-8845 x100 • Fax: 717-533-8661
E-Mail: cust@igi-global.com • www.igi-global.com

Editorial Advisory Board

Table of Contents

Detailed Table of Contents

This chapter extends the discussion started in the first chapter by explaining how green considerations can be applied during all phases of the public procurement process: from the pre-award to the award and post-award phase. It presents how green provisions can be included in technical specifications, award criteria and contract performance clauses. The chapter points the lacking appropriate regulations at national and international level, or in the actual lack of training of the procurement workforce.

This chapter touches upon the Green Audit that works as an environmental consultancy and review organization with an aim of monitoring the performance of companies and organizations whose activities might threaten the environment and the health of citizens. The chapter suggests that a network of communication system need to be created with similar organizations and environmental databases worldwide so that any new threat to humanity can be monitored and evaluated.

The chapter is based on a study that aims to highlight the advantages of GPP practices to address the environmental and economic problems, laying on the experiences of the European countries. The research has the objective to reconstruct the state of the art of green public procurement through the lens of a cross-country comparative analysis at European level. After introducing a systematic review of the literature and the core regulations of the GPP practice, the chapter underlines the results of a multidimensional analysis on a cluster of 80 GPP practices (sample updated to February 2014), identified in European Union and implemented by governments in 25 countries at a central, regional and local government level. The framework of analysis builds on several dimensions, mapping the main results on the following levels: geographic origin, the government level, GPP implementation period, main criteria for implementation, as well as environmental and economic impact of such practices.

The seventh chapter is a chapter on the initiations taken forward by the member states of East African Community (EAC) on green public procurement implementation in the states of EAC. The further discusses on country specific status and challenges, provides issues on implementation and monitoring, recommendations on performance indicators and action activities for the way forward.

This chapter presents a case study of Bhutan in implementing green public procurement, which is one of the most successful green public procurement programs implemented in a developing country. The chapter could be a guide for the emerging countries to follow for the implementation of GPP as the chapter covers the approaches of implementation, a checklist of key considerations, and a comprehensive list of valuable lessons learned.

Preface

Green public procurement (GPP) is a public procurement, which integrates environmental considerations in the public procurement process and takes into the account the entire lifecycle of the product or service. Also, it is considered to be appropriate and one of the most useful tools seeking to reduce negative impacts on the environment, promoting social responsibility, and contributing to the economic development (Shakya, 2016).

Public procurement is the core activity of all sorts of development policy implementations. Government consumption accounts for a sizeable part of economic activity and demand. UN Environment DTIE (2012) revealed the statistics of public spending concerning GDP. The share of public procurement in OECS countries annually accounts for an average of 12% and up to 30% in developing countries. The annual public-sector procurement budget provides the public sector with considerable leverage to stimulate the public-sector consumptions favoring goods, works and services using fewer resources, and less harmful to the environment. Better policies and strategies on green procurement and sustainable use of resources help favorable impact on our environment and life. Much of the spending by public authorities are in sectors with high environmental impacts, such as transport, buildings, and furnishings. As purchasers, public authorities have massive buying power, and there is enormous scope for them to influence suppliers to innovate and produce more environmentally friendly goods and services. The public sector also has the potential to persuade private sector companies and the general public to change their consumption habits.

The way humans consume resources in the world is causing environmental damage at a rate that cannot be sustained. Green public procurement (GPP) is widely recognized internationally as an effective means for public administrations to manage the balance between cost-effectiveness and sustainable development. Citizens need to be sure that what is procured on their behalf is procured in such a way as to eliminate harmful effects on the environment and society. At the same time, these goods, services, and works must represent both short-term and long-term value for money.

Social responsibility of public procurement is about procuring entities taking into consideration the societal impacts of their purchases in a broader sense than merely looking at the purchasing price or the price-quality relationship. The guiding influence of public procurement, which is of crucial importance in society, is associated with corporate social responsibility. By promoting socially responsible public purchases, the entities can provide bidding community (suppliers, contractors, consultants) with substantial incentives to develop corporate social responsibility in their activities.

This book looks to discuss and address the challenges that different countries have faced in because of climate change, reasons of climate change, how public procurement can contribute to improving the world environment through green public procurement (GPP). It will also discuss how green public procurement can be incorporated seamlessly in the procurement processes.

This book aims to be an essential practical reference source, building on the available literature in the field of green public procurement (GPP) and providing for further research opportunities. It is hoped that this text will provide the resources necessary for policymakers, practitioners, development agencies and suppliers to adopt and implement green public procurement for the sustainable procurement practice and contributing towards the better green world.

This book is published for its scholarly value and contributions to the field of green public procurement (GPP), providing for further research opportunities which could lead to broader incorporation of green procurement in public procurement practices and procedures in a seamless manner.

Policy makers, academicians, researchers, advanced-level students, Suppliers, development agencies, and government officials will find this book useful in furthering their procurement practice, research exposure to pertinent topics in green public procurement (GPP) and Sustainable Public Procurement, developing green public procurement strategies and policies, and assisting in furthering their own research efforts in this field.

The topics discussed in the book include different aspects of sustainable public procurement (SPP) and green public procurement (GPP) covering the unique compilation of best practices, policy guidance, research findings, practitioner-oriented green public procurement implementation case studies from two countries, which provide a holistic body of knowledge on Green Public Procurement.

AN OVERVIEW OF CHAPTERS

The first three chapters of this book provides different aspects of green public procurement (GPP) in terms of its policy and process issues, Chapters 4 and 5 are fully based on research, and Chapters 5, 7, and 8 are the case studies from Asia, Africa, and Latin America and the Caribbean (LAC).

The first chapter of the book is titled "Green Public Procurement Implementation Framework: Strategies for Environmental Sustainability," which introduces the basics of green public procurement (GPP) and sets a tone for the whole book taking account of emerging international trends whereby the sustainability of procurement policy is gauged regarding social and economic as well as environmental impacts. The chapter reviews the current challenges in policy and strategy development, provides guidance based on the established best practices in the Europe and other parts of the world for developing policies and strategies, presents the ways to incorporate GPP in the traditional procurement processes and practices.

The second chapter, titled "Implementation of Green Considerations in Public Procurement: A Means to Promote Sustainable Development," extends the discussion started in the first chapter by explaining how green considerations can be applied during all phases of the public procurement process: from the pre-award to the award and post-award phase. Presents how green provisions can be included in technical specifications, award criteria and contract performance clauses. The chapter points the lacking appropriate regulations at national and international level, or in the actual lack of training of the procurement workforce.

The third chapter, "Green Audit for the Environmental Sustainability," touches upon the Green Audit that works as an environmental consultancy and review organization with an aim of monitoring the performance of companies and organizations whose activities might threaten the environment and the health of citizens. The chapter suggests that a network of communication system need to be created with similar organizations and environmental databases worldwide so that any new threat to humanity can be monitored and evaluated.

The fourth chapter is based on the research on the documental analysis of the 80 green public procurement practices in European Union, and provides mapping of the principal green public procurement practices in Europe and their environmental impact. The researchers present a map of public administrations initiatives and their critical reading on the basis of the main theories and models considered.

The fifth chapter is a research-based and provides a perception of green production in the textile industry for environmental sustainability through SWOT and AHP analysis. The chapter provides the study outcomes on investigation and identification of the most appropriate strategy about waste management for the textile sector in Turkey. Also, this study provides a point of general view and approach on the importance of green production and consumption in the Textile Industry.

The sixth chapter provides a case study of the strategic implementation of green public procurement in the City of Buenos Aires. As public procurement systems around the world continue their modernization process, many countries are including a new objective in their reform efforts: the implementation of public policies in the procurement process. This has been a growing trend in Latin America, where countries have shown leadership in incorporating public policies into their modernization plans, being a perfect example is the introduction of green public procurement standards. There is particular interest in incorporating sustainability in public procurement at the sub-national level where governments are directly responsible for the provision of citizen services. In the case of Argentina, the Government of the City of Buenos Aires is at the forefront of public procurement modernization, leading the country in green public procurement strategies. This chapter explores the principles and global trends that have influenced the City of Buenos Aires in implementing green public procurement and provides insight on the economic rationale and potential benefits for one of the largest metropolitan regions in LAC.

The seventh chapter is a chapter on the initiations taken forward by the member states of East African Community (EAC) on green public procurement implementation in the states of East African Community (EAC). The further discusses on country specific status and challenges, provides issues on implementation and monitoring, recommendations on performance indicators and action activities for the way forward.

Chapter 8 presents a case study of Bhutan in implementing green public procurement, which is one of the most successful green public procurement programs implemented in a developing country. The chapter could be a guide for the emerging countries to follow for the implementation of GPP as the chapter covers the approaches of implementation, a checklist of key considerations, and a comprehensive list of valuable lessons learned.

The book as a whole discusses on the green public procurement (GPP) policy issues and their outcomes to achieve better environment friendly, socially responsible and economically sustainable procurement system for the better life on earth.

Rajesh Kumar Shakya
The World Bank, USA

REFERENCES

Shakya, R. K. (2016). *Green Public Procurement Strategy for East African Community (EAC)*. Arusha, Tanzania: EAC.

Acknowledgment

I thank my wife, Dr. Dhandu Rani Shakya, my son, Ayush Shakya, and daughter, Ayusha Shakya, for the patience and encouragement on the development of this book. Without their consistent support, this would have been impossible.

I want to thank and acknowledge the work of each of the authors that contributed the book. I thank their dedication, their patience, and their ideas that led this book to be a pillar of knowledge in e-Government Procurement (e-GP).

I thank my Editorial Advisory Board, who reviewed the chapters and provided comments and valuable suggestions for enhancing the quality of the book.

Rajesh Kumar Shakya
The World Bank, USA

Chapter 1
Green Public Procurement Implementation Framework:
Strategies for Environmental Sustainability

Rajesh Kumar Shakya
The World Bank, USA

ABSTRACT

Government expenditure on public procurement accounts for a sizeable part of economic activity and demand. The annual public-sector procurement budget accounts about 20% to 30% of GDP in the developing countries, which provides the countries public sector with considerable leverage to stimulate the public-sector consumptions favoring goods, works, and services using fewer resources, and less harmful to the environment. The public sector should be more responsible for supporting environmental and broader sustainable development objectives. Green public procurement (GPP) is slowly getting recognized internationally as an effective means for public administrations to manage the balance between cost-effectiveness and sustainable development. The public fund should be used responsibly to uplift the health, environment, and social life. At the same time, the goods, services, and works procured by the government must achieve value for money. This chapter focuses on the importance of green public procurement strategies that could guide the countries to successfully plan, procure, and implement green public procurement (GPP) by outlining policies, strategies, and actions to boost green public procurement. The recommended strategies are based on the international best practices and the lessons learned from the experience by different government around the world. These strategies could be updated and customized based on the specific country context and their readiness.

DOI: 10.4018/978-1-5225-7083-7.ch001

INTRODUCTION

Public procurement is the core activity of all sorts of development policy implementations. Government consumption accounts for a sizeable part of economic activity and demand. UN Environment DTIE (2012) revealed the statistics of public spending in terms of GDP. The share of public procurement in OECS countries annually accounts for an average of 12% and up to 30% in developing countries. The annual public-sector procurement budget provides the public sector with considerable leverage to stimulate the public-sector consumptions favoring goods, works and services using fewer resources, and less harmful to the environment. Better policies and strategies on green procurement and sustainable use of resources help favorable impact on our environment and life. The growing initiatives for green procurement by countries around the world is in support of the goals of the United Nations (UN), which is to become climate neutral and environmentally sustainable. This chapter sets a tone for the whole book and takes account of basics of green procurement and emerging international trends whereby the sustainability of procurement policy is gauged regarding social and economic as well as environmental impacts.

DEFINITIONS OF GREEN AND SUSTAINABLE PROCUREMENT

Different pieces of literature and governments have presented several definitions on the Sustainable Public Procurement (SPP) as well as on the Green Public Procurement (GPP) but primarily focused on the process. A set of such practical definitions are presented here.

Sustainable Public Procurement (SPP) is a process whereby organizations meet their needs for goods, services, works and utilities in a way that achieves value for money on a whole life basis for the benefit of society and the economy (Indiana University, 2018), while minimizing damage to the environment. SPP has three pillars, viz.: economic development; social development; and environmental protection (Figure 1).

Green public procurement (GPP) is a public procurement, which integrates environmental considerations in the public procurement process and takes into the account the entire lifecycle of the product or service. Also, it is considered to be appropriate and one of the most useful tools seeking to reduce negative impacts on the environment (Shakya, 2016).

Moreover, GPP is a process whereby public agencies fulfill their operational needs of goods, works, services, and utilities by soliciting and selecting products and services that have less influence on the environment during their expected economically - useful life, as compared to alternative products/solutions. Prices of the goods, works,

Figure 1. The three pillars of sustainable development
Source: UNEP, 2011

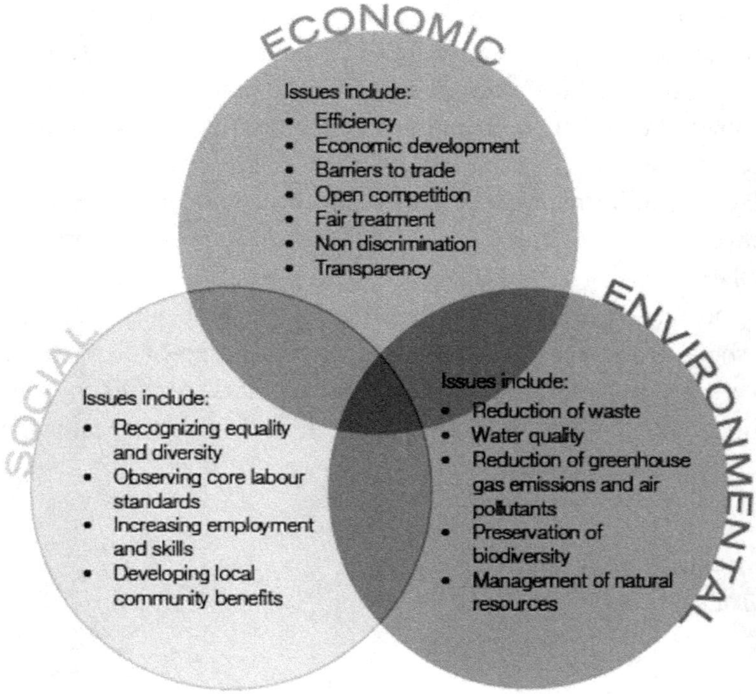

and services should consider environmental costs. GPP involves the synthesis of ecological criteria into public procurement processes. In specifying such criteria, procuring entities encourage the promotion of environment-friendly technologies and the development of environmentally sound products. Where possible, public procurement policies should promote the use of all types of environment-friendly products, which allows using energy, water, non-toxic, low emission chemicals.

The scale of public procurement offers a powerful policy instrument alongside more traditional instruments such as regulation, policy direction, and economic mechanisms like taxation, to control the market in favor supply of 'green' goods and services.

GREEN PUBLIC PROCUREMENT STRATEGIES

The aim of establishing Green Public Procurement strategies is to promote GPP and ensure that the purchased goods and services would be as environmentally friendly as possible and help human health better by making our planet green and pollution free for human and other living creatures. Guided by the UN Sustainable Development Goals, countries seem to be mostly trying to align their different policies including the public procurement policy and practice consider the strategies for promoting environmental aspects (green and sustainable public procurement) and adhering to specific social and ethical standards (socially responsible public procurement). Figure 2 shows a systematic steps for the development of green or sustainable green procurement strategies.

Social responsibility of public procurement is about procuring entities taking into consideration the societal impacts of the purchases in a broader sense than merely looking at the purchasing price or the price-quality relationship. The guiding influence of public procurement, which is of crucial importance in society, is associated with corporate social responsibility. By promoting socially responsible public purchases, the entities can provide bidding community (suppliers, contractors, consultants) with tangible and intangible incentives to develop corporate social responsibility in their supply-chain activities.

Figure 2. Developing a green, or sustainable, public procurement strategies
Source: UNESCAPE, 2012.

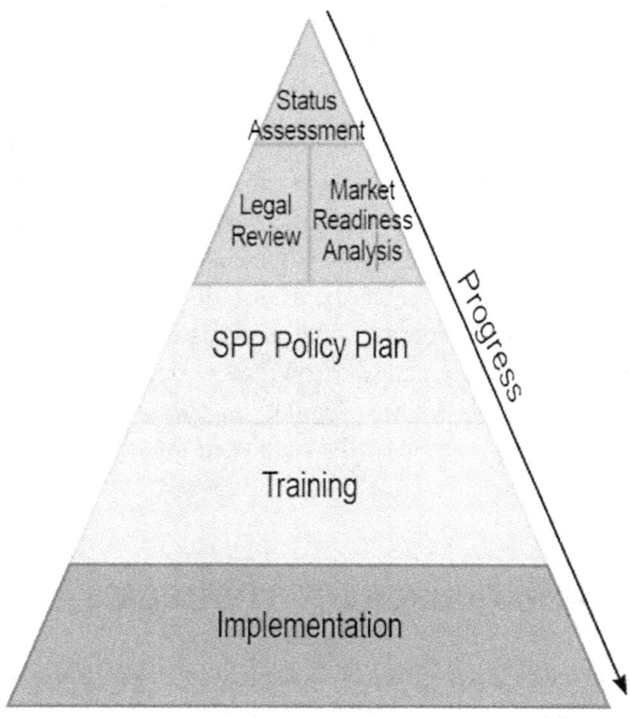

Countries with newly adopted procurement legislations are more proactive to the sustainable procurement and green public procurement. For example, Kosovo has prepared its Green Procurement strategy and identified its own set of strategic items with desired green standards. PPRC-Kosovo (2014) recommends, "Successful implementation of Green Procurement requires the identification and implementation of environmental performance opportunities at both the strategic and operational levels, taking into consideration specific buying patterns, sustainable development targets, and other government priorities (p. 35)."

Public procurement has an immediate impact on the economy. The influence on controlling the price, access, and the quality of goods and services used in construction, cleaning, irrigation, health, and other sectors are highly dependent on the government policy on the environment, applicable standards, and policy practice during the procurement process.

PPRC-Kosovo (2014) further highlights, that the green public procurement attempts to decrease the environmental consequences of government activities and promote environmental stewardship by integrating environmental performance considerations in the procurement process as part of an ongoing commitment to improving the environment and the quality of life.

In public procurement, Green procurement should still satisfy the need of value for money, buy appropriately including environmental performance into the procurement lifecycle processes including planning, acquisition, consumption, and disposal. Value for money, in addition to the economy, efficiency, and effectiveness, should be factored by environmental performance and relevant associated costs with the life cycle of procured goods and services. Also, the supporting administrative processes and procurement methods can also offer opportunities to reduce the environmental impacts of government operations.

STRATEGIC OBJECTIVES

European Union has developed a more comprehensive green public procurement strategy and has outlined its strategic objectives in two broader categories, general and specific. A similar division of objectives sounds reasonable as the market for public procurement usually revolves around the national and regional boundaries. General objectives of Green Public Procurement are to:

- Increase the level of Green Public Procurement use at the national and local level, and
- To create demand for products that meet high environmental standards, and innovative environmental technologies by the government entities.

Moreover specific objectives are to:

- Increase knowledge and experience of the use of Green Public Procurement;
- Increase in the rate of procurements taking into account environmental aspects, and
- Increase the number of Suppliers holding a verified environmental management system and producing ecologically certified products.

As far as the social responsibility of public procurement is concerned, the primary objective is to disseminate the strategic perspective of public procurement building on awareness and knowledge.

Specific objectives are to:

- Encourage procuring entities and suppliers to cooperate not only in contracts but mainly by partnerships in green initiatives,
- Initiate a consultation process during the preparation of procurement activities, using "bottom-up" meetings, focus groups at the local and regional level and professional expert panels for the use of green products;
- Spread the idea of socially responsible green public procurement by open access to information about calls, tenders, and contracts;

STRATEGY STATEMENTS

To achieve this objective, the governments may:

1. Initiate and stimulate debate regarding Green Public Procurement (GPP) and Sustainable Public Procurement (SPP) in the country;
2. Exchange of good practice between procuring entities by establishing a platform for sharing experiences, information, and knowledge on the adaptations for GPP and SPP;
3. Elaborate concrete advice, criteria, and indicators to be used in public tendering given promoting GPP and SPP; and
4. Foster public procurement planning, to timely, integrate GPP and SPP criteria into the procurement process.

The green public procurement as a public procurement whereby the Procuring Entity includes into the bidding documents at least the minimum environmental standards for the procurement of goods, works, and services in addition to their price and quality throughout their life-cycle (Šličiuvienė, 2017). thereby stimulating the production of environment-friendly products. Green procurement enables the procuring entity to acquire the product which (if compared to the product of the same function):

- Consumes less natural resources or uses them more sustainably for its production and consumption and therefore pollutes less;
- Is made from renewable, ecological sources;
- Contains fewer or none hazardous, toxic materials;
- Is durable, functional and does not have a negative impact on the humans‘ health and environment;
- Can repeatedly be used;
- At the end of its use can be reused or recycled.

The promotion of GPP is one of the policy interventions and can serve as a significant catalyst for change in environmental protection efforts. Public procurement can shape production and consumption trends and open up a new market for green products.

GPP is also an opportunity for innovation and competitiveness. Small and Medium Enterprises (SME) can benefit from this opportunity to enter into a new market and pioneer in the niche market. The process of socio-economic transformation presently underway in most of the developing countries will provide such opportunities and help SMEs achieve competitive advantages.

ACTION AND SUPPORT MEASURES NEED TO BE TAKEN

While setting an ambitious target of reaching the level required in GPP application, governments may need to outline, among other things, a variety of actions and support measures. For example, such measures include:

- A regular supplement of the list of products for which the environmental criteria are to be applied (Šličiuvienė, 2017);
- Support for civil society organizations which actively contribute to the promotion of the environmentally-friendly (green) public procurement and consumption of organic products (Siemens, 2003);

- Preparation and distribution for the procuring entities of the guidelines and publications about the green public procurement;
- Installation of environmental management systems or standards in public institutions;
- Periodical improvement of the green procurement training program and organization of the training for the procurement officials.

There is a general perception that green public procurement are expensive. However, if the full lifecycle costs (purchase price, usage, disposal costs) of a contract considered and not only the purchase price, the cost spent on greener products are justified. The PricewaterhouseCoopers, Significant and Ecofys indicated in the 2009 Report on Collection of Statistical Information on Green Public Procurement in the EU that GPP can lead to decreases in costs for the procuring entity instead of increases. The report showed that "when using a Life Cycle Costing (LCC) approach in calculating the financial impact of GPP, the outcome is that with an average level of GPP of 45%, the average financial impact of GPP is -1%" (PricewaterhouseCoopers, Significant and Ecofys, 2009). The green procurement can decrease the overall average cost by 1% even though the initial purchase cost may seem higher. The lower operating price because of lifecycle consideration compensates the higher initial value. The other study also supported this conclusion by stating that Lower life-cycle costs of GPP-compliant goods and services and increasing competition over time as GPP compliance initially gives a competitive advantage are assumed to restore the balance of the prices of products and services purchased with GPP requirements.

Furthermore, it was concluded in the before mentioned 2009 report (PricewaterhouseCoopers, Significant and Ecofys) that there are mainly two product groups leading to cost reductions through GPP: construction and transport.

Šličiuvienė (2017) also identified numerous other barriers and concerns directly associated with the adaptation of green public procurement. These are, for instance:

- **Support From Senior Management:** Because of the existing perception that GPP costs more and the annual budget constraints on the public entity, there could be a lack of support for green public procurement from senior management.
- **Consultation With Suppliers:** Consultation between the Procuring Entities, suppliers, environmental experts and the end users of the products should be encouraged to enhance the green public procurement. However, the public agencies are usually do not engage with the suppliers during the pre-procurement consultation stage because of the strict legislation regulating the preferential treatment of the suppliers.

- **Information, Knowledge, and Practical Experience:** Many purchasers within public agencies often lack the information concerning the more economical alternatives, also understanding of environmental criteria and practical experience on the application of green public procurement. Consequently, the inefficient purchasing decisions that not only hurt the environment but also waste government resources. Also, Green Public Procurement is seen as a more complex, detailed and resource intensive process which places an enormous burden on purchasers. Therefore, intensive training on the legal and technical aspects of GPP implementation is necessary (pp. 26-27).

ENVIRONMENTAL REQUIREMENTS INTO PROCUREMENT PROCESS

Procuring Entities must obtain the best value for money through competition among the suppliers, and act equitably in the course of public procurement (apply the principles of non-discrimination, equal treatment, transparency, and proportionality). Public agencies should be careful to balance primary procurement objectives (i.e., delivering goods and services necessary to accomplish government mission in a timely, economical and efficient manner) with secondary policy objectives (such as promoting sustainable green procurement and promoting local content). Such an inclusive policy brings a balance between financial and environmental, and sustainable procurement benefits. The Green Public Procurement practices implementation will be evident through an integrated approach and including environmental criteria at every stage of procurement procedure. The procedure should incorporate standard specifications of design to prequalified and potential suppliers, environmental management systems (like the concept of *3Rs: Reduce, Reuse and Recycle*) being introduced through waste management, product contents restrictions, checking supplier environment objectives and vision, and vendor auditing on compliance.

INCLUSION OF ENVIRONMENTAL CRITERIA UNDER STATE LAW

There are some ways under the state procurement laws, which can provide the opportunity for the procuring entities to incorporate environmental requirements into the procurement process. State procurement law would need agreeing to subject its procurement procedures to a certain degree of international and regional regulation, which establishes some rules and principles, which must be observed in the award of the public contracts. The countries can develop and adopt the policy framework which guides and assists in implementing procurement policies that seek to address not only economic but also environmental objectives. Environmental policy and strategy should cover all the areas with the ecological impact to contribute to the preservation, safety, and improvement of the quality of the environmental problems.

Recent amendments to public procurement laws and new public procurement policy documents in many countries reflect the aspirations for sustainable development by protection and improvement of the quality of the environment. Following flowchart (Figure 3) shows the procurement process with indication of sustainability and green interventions:

Requirements of Green Public Procurement (GPP) should be incorporated in the public procurement through the amendment in regulations or separate government guidelines on Green Public Procurement. *The regulation or guidelines should contain some provisions that refer specifically to the incorporation of the environmental considerations into the bidding procedures.* If a procuring entity wishes to include ecological requirements, it has a choice of approaches listed below.

1. **Technical Specifications:** The procuring entities may include a minimum level of environmental performance in the technical requirements of goods, services or works being purchased if it contributes to the fulfillment of their use. Technical specifications may be formulated regarding performance or functional requirements and by reference to international, national or regional standards or equivalent, as well as to specific production process or method provided they be linked to the contract, proportionate to its value and objectives, and non-discriminatory. Technical specifications shall afford equal access of economic operators to the procurement procedure and shall not have the effect of unjustified obstacles to the opening up of public procurement to competition. Also, characteristics required of a material, product, supply or service may also include levels of environmental and climate performance. Importantly, as an ability to meet the technical specifications is a prerequisite for being considered a candidate for the contract, the environmental demands included therein is compulsory.

The Law on Public Procurement requires reiteration of the provisions and provides that without bias to mandatory national technical standards, to the extent that they are compatible (The Public Contracts Regulations, 2015) with regional and international requirements, the technical specifications may be formulated regarding performance or functional elements of the subject-matter of the contract. Functional requirements may include environmental characteristics. Such conditions must be sufficiently precise to allow suppliers to determine the subject-matter of the

Figure 3. The procurement process with indication of sustainability and green interventions
Source: UNEP, 2011

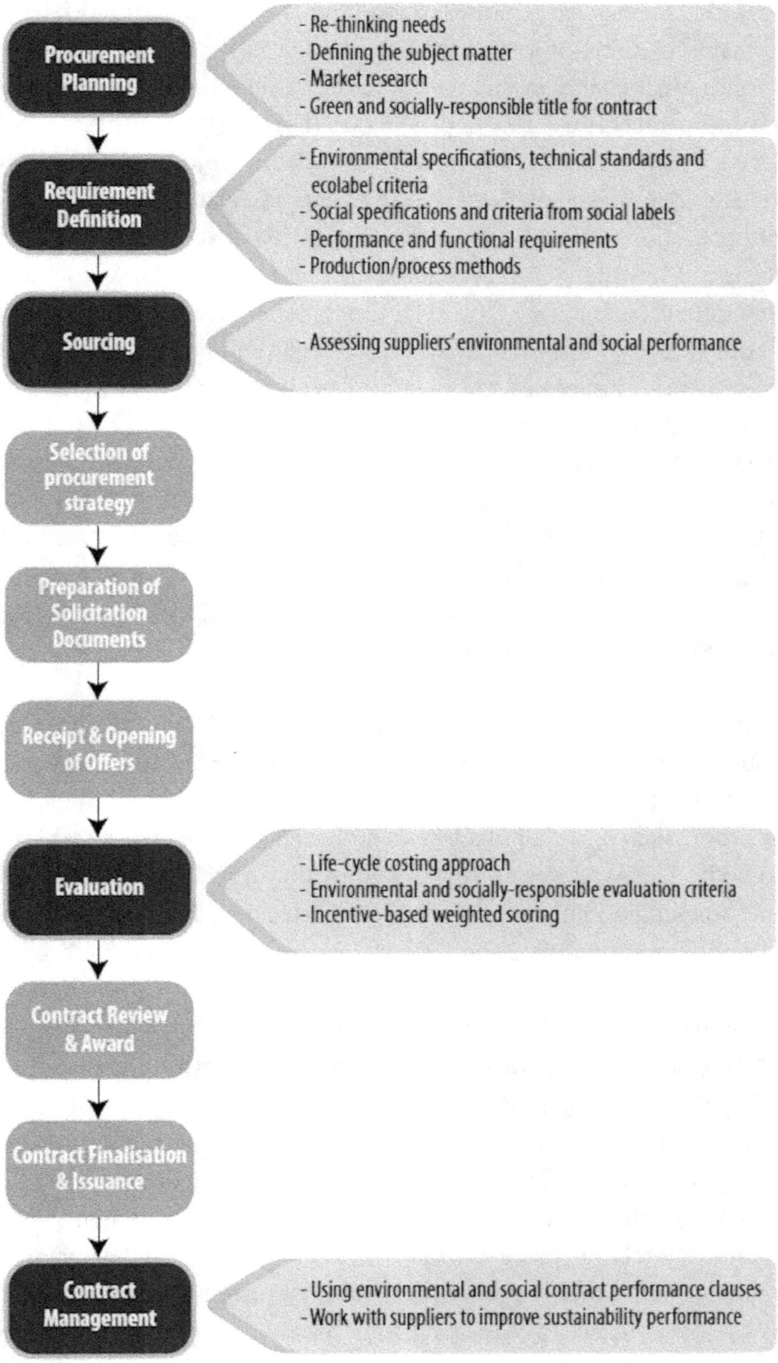

contract and to facilitate procuring entities to award the contract for the necessary supplies, services or works.

2. **Eco-Labels:** Procuring Entities can use eco-labels, regional eco-labels, (multi) national eco-labels or any other label which provide information about the environmental characteristics of the products and thus allow to compare the environmental performance of products of the same type. The eco-labels can be used if the requirements for them are linked to the subject-matter of the contract and meet specific standards of openness and transparency. A certain eco-label cannot be required because it would be considered discriminatory. However, the criteria used by that eco-label can be included in the bidding documents. The public procurement law in this regard may state that where a procuring entity lays down environmental characteristics regarding performance or functional requirements, it may use the detailed specifications, or, if necessary, parts thereof, as defined by regional or (multi-) national eco-labels, or by any other eco-label. Those specifications are appropriate to describe the characteristics of the products or services that are the object of the contract. The requirements for the eco-label can be drawn upon the basis of scientific information. The eco-labels are adopted using a procedure in which government organizations, purchasers, manufacturers, distributors, environmental organizations and other interested persons can participate.

The procuring entity may also indicate that the products and services bearing the eco-label are considered to address the technical specifications laid down in the technical specification documents. In such a case, it must accept any other appropriate means of proof, such as a manufacturer's authorization letter or a test report from a notified body. The most extensive global directory of ecolabels "Ecolabel Index" is currently tracking 463 ecolabels in 199 countries, and 25 industry sectors. Eco-labels could be for agricultural and food products, non-food products, etc.

3. **Exclusion of Supplier on the Ground of Non-Compliance With Environmental Legislation:** The procuring entity may exclude from participation in procurement procedure any Supplier, which has violated the applicable environmental obligations. A case of non-compliance with environmental legislation could be treated as professional misconduct under the state law, and the bidders may be excluded from the procurement procedures if this misconduct took place less than one year ago. However, decisions to reject candidates are should be proportionate to the offense and that the candidate gets the opportunity to take corrective measures or efforts they might have made to prevent recurrence.

4. **Environmental Technical Competence:** Procuring entities can impose requirements ensuring that Suppliers possess the environmental technical expertise necessary to fulfill the contract, e.g., have access to the professional equipment for environmental protection.
5. **Environmental Management Systems or Standards:** Public purchasers can require Suppliers to demonstrate that they comply with environmental management systems or standards, which are relevant to the performance of the contract in question (for service and works contracts only) and which certify that companies operate to certain levels of environmental standards and awareness. An organization running an environmental management system may request certification under one of the two principal environmental management systems in use: the Eco-management and audit process or the international standard on environmental management systems (EN/ISO 14001). Companies may also be able to demonstrate that they apply equivalent environmental management measures, even without certification. About environmental management systems, the law or regulation may state that the procuring entities, in public works and service contracts, require the certificates issued by independent authorities attesting compliance of the supplier with some environmental management standards, must refer in the contract documents or to environmental management standards. The standards must be certified by authorities conforming to national, regional, international standards certification. The contracting authorities shall recognize equivalent certificates from bodies established in other countries. They may also accept other evidence of comparable environmental management models from the suppliers."
6. **Contract Award Criteria:** The application of the contract award criteria based on environmental characteristics is another possible choice. Award criteria enable the purchasing authority to score each bid by its satisfaction of different combinations of rules, including those related to the environmental protection. The additional points can be awarded to the candidates that go beyond minimum requirements thereby stimulating greater sustainability. It is crucial to ensure, however, that the assessment, which has taken place at the selection stage, is not duplicated at the award stage. Also, the choice of environmental qualification and award criteria must respect the single market principles which serve to uphold competition: equal treatment, transparency, non-discrimination, proportionality, and mutual recognition. In such situations, the pre-procurement consultation can be beneficial when formulating more appropriate and ambitious award criteria related to environmental objectives.

However, as it was mentioned earlier, procuring entities are discouraged from pre-procurement consultation because of the strict legislation regulating the preferential treatment of the suppliers. Suppliers, in turn, is uncertain whether they could take part in the further procurement process if involved in the pre-procurement consultation.

7. **Life-Cycle Costing:** Contracting authorities can apply the life-cycle costing to measure and compare costs including those attributed to environmental externalities.
8. **Contract Performance Clauses:** Environmental conditions for the performance of a public contract can be set up provided that they be linked to the contract and indicated in the Invitation to Bid (ITB) or in the procurement documents. In this case, the indication of such clauses in the procurement documents is a prerequisite to the equal treatment of tenderers and transparency. Unlike contract award criteria, which are the basis for a comparative assessment of the quality of bids, contract performance conditions constitute fixed requirements that have no impact on the evaluation of proposals. However, having environmental contract performance clauses is only useful if compliance requirements are regularly and impartially monitored. Because of the lack of human resources, also the difficulties of verifying compliance with all requirements, it could be a rare practice in many countries. Consequently, an authorized person from the procuring entity who can sometimes be biased or lack the relevant competence is usually trusted with this function.

There is a general practice while introducing GPP in the procurement process. Usually, countries develop core and broad criteria for a series of product and service groups, which are regularly reviewed and updated. They are designed to be inserted directly into the Standard Bidding Documents (SBDs) and thereby help to procure authorities to reach a right balance between environmental performance, cost considerations, market availability and ease of verification.

Two sets of criteria are prepared for each of the product and service group:

- **Core Criteria:** these are the common criteria designed to be used by any Procuring Entity. The criteria should meet the most significant environmental outcomes and should be used with minimum additional verification effort or cost increases.
- **Broad Criteria:** Any procuring entity should use these criteria for buying the best products complying with environmental standards available on the market. Procuring entities may need to carry out additional verification to fulfill this criterion, and it may increase a little more cost that complying with the core criteria.

European Union has prepared a comprehensive set of criteria for different products and their categories. Countries may refer and use the European Commission GPP criteria as a basis for their own GPP criteria (http://ec.europa.eu/environment/gpp/eu_gpp_criteria_en.htm) than developing everything from scratch on their own (European Commission, 2016).

GREEN PUBLIC PROCUREMENT POLICY PROVISIONS

The Policy Provision

The Environmental Performance Index (EPI) Report (2018) by Yale and Columbia University that scores 180 countries on 24 performance indicators across ten categories states that careful measurement of environmental trends and progress provides a foundation for effective policymaking. The metrics provide a gauge at a national scale of how close the countries are to reach environmental policy goals. Despite efforts of the governmnets, it illustrates how the issue of environmental sustainability in most of the countries is still at the very low level (Figure 4).

It is recognized that the implementation of policies to advance Sustainable procurement and GPP takes place in the context of the rules and legislation dealing with public procurement. As previously stated, the extent to which environmental and social criteria may be included in bidding documents and contracts for public sector procurement. These regulations provide that criteria adopted by procuring entities contracting by most economically-advantageous tender (MEAT) shall be linked to the contract and may include environmental characteristics. Also, the weight of which marks are to be allocated must be stated. If environmental characteristics are specified, this must be done with sufficient precision to determine the subject matter of the contract and to allow the contract to be awarded.

In certain circumstances, the legislation should allow for procuring entities to require their suppliers to hold certification in respect of environmental management systems. Again, any such stipulation must be related to the subject matter of the contract. In this case, the procuring entity shall refer either to the Eco-Management and Audit Scheme (EMAS) or other international environmental management standards. Other environmental management measures that can be proved to be equivalent must be accepted.

Life Cycle Approach

Green Public Procurement (GPP) will be effective only through the life-cycle thinking in procurements. Analysis on this basis enables public procurers, their

Figure 4. The 2018 EPI Rankings. Rank, EPI score, and regional standing for 180 countries
Source: EPI, 2018

2018 EPI RANKINGS

RANK	COUNTRY	SCORE	REG
1	Switzerland	87.42	1
2	France	83.95	2
3	Denmark	81.60	3
4	Malta	80.90	4
5	Sweden	80.51	5
6	United Kingdom	79.89	6
7	Luxembourg	79.12	7
8	Austria	78.97	8
9	Ireland	78.77	9
10	Finland	78.64	10
11	Iceland	78.57	11
12	Spain	78.39	12
13	Germany	78.37	13
14	Norway	77.49	14
15	Belgium	77.38	15
16	Italy	76.96	16
17	New Zealand	75.96	1
18	Netherlands	75.46	17
19	Israel	75.01	1
20	Japan	74.69	1
21	Australia	74.12	2
22	Greece	73.60	18
23	Taiwan	72.84	2
24	Cyprus	72.60	19
25	Canada	72.18	20
26	Portugal	71.91	21
27	United States of America	71.19	22
28	Slovakia	70.60	1
29	Lithuania	69.33	2
30	Bulgaria	67.85	3
30	Costa Rica	67.85	1
32	Qatar	67.80	2
33	Czech Republic	67.68	4
34	Slovenia	67.57	5
35	Trinidad and Tobago	67.36	1
36	St. Vincent & Grenadines	66.48	2
37	Latvia	66.12	6
38	Turkmenistan	66.10	7
39	Seychelles	66.02	1
40	Albania	65.46	8
41	Croatia	65.45	9
42	Colombia	65.22	2
43	Hungary	65.01	10
44	Belarus	64.98	11
45	Romania	64.78	12
46	Dominican Republic	64.71	3
47	Uruguay	64.65	3
48	Estonia	64.31	13
49	Singapore	64.23	3
50	Poland	64.11	14
51	Venezuela	63.89	4
52	Russia	63.79	15
53	Brunei Darussalam	63.57	4
54	Morocco	63.47	3
55	Cuba	63.42	4
56	Panama	62.71	5
57	Tonga	62.49	3
58	Tunisia	62.35	4
59	Azerbaijan	62.33	16
60	South Korea	62.30	5
61	Kuwait	62.28	5
62	Jordan	62.20	6
63	Armenia	62.07	17
64	Peru	61.92	6
65	Montenegro	61.33	18
66	Egypt	61.21	7
67	Lebanon	61.08	8
68	Macedonia	61.06	19
69	Brazil	60.70	7
70	Sri Lanka	60.61	6
71	Equatorial Guinea	60.40	2
72	Mexico	59.69	8
73	Dominica	59.38	5
74	Argentina	59.30	9
75	Malaysia	59.22	7
76	Antigua and Barbuda	59.18	6
77	United Arab Emirates	58.90	9
78	Jamaica	58.58	7
79	Namibia	58.46	3
80	Iran	58.16	10
81	Belize	57.79	10
82	Philippines	57.65	8
83	Mongolia	57.51	9
84	Serbia	57.49	20
84	Chile	57.49	11
86	Saudi Arabia	57.47	11
87	Ecuador	57.42	12
88	Algeria	57.18	12
89	Cabo Verde	56.94	4
90	Mauritius	56.63	5
91	Saint Lucia	56.18	8
92	Bolivia	55.98	13
93	Barbados	55.76	9
94	Georgia	55.69	21
95	Kiribati	55.26	4
96	Bahrain	55.15	13
97	Nicaragua	55.04	14
98	Bahamas	54.99	10
99	Kyrgyzstan	54.86	22
100	Nigeria	54.76	6
101	Kazakhstan	54.56	23
102	Samoa	54.50	5
103	Suriname	54.20	15
104	São Tomé and Príncipe	54.01	7
105	Paraguay	53.93	16
106	El Salvador	53.91	17
107	Fiji	53.09	6
108	Turkey	52.96	24
109	Ukraine	52.87	25
110	Guatemala	52.33	18
111	Maldives	52.14	10
112	Moldova	51.97	26
113	Botswana	51.70	8
114	Honduras	51.51	19
115	Sudan	51.49	14
116	Oman	51.32	15
117	Zambia	50.97	9
118	Grenada	50.93	11
119	Tanzania	50.83	10
120	China	50.74	11
121	Thailand	49.88	12
122	Micronesia	49.80	13
123	Libya	49.79	16
124	Ghana	49.66	11
125	Timor-Leste	49.54	14
126	Senegal	49.52	12
127	Malawi	49.21	13
128	Guyana	47.93	20
129	Tajikistan	47.85	27
130	Kenya	47.25	14
131	Bhutan	47.22	15
132	Viet Nam	46.96	16
133	Indonesia	46.92	17
134	Guinea	46.62	15
135	Mozambique	46.37	16
136	Uzbekistan	45.88	28
137	Chad	45.34	17
138	Myanmar	45.32	18
139	Côte d'Ivoire	45.25	18
140	Gabon	45.05	19
141	Ethiopia	44.78	20
142	South Africa	44.73	21
143	Guinea-Bissau	44.67	22
144	Vanuatu	44.55	7
145	Uganda	44.28	23
146	Comoros	44.24	24
147	Mali	43.71	25
148	Rwanda	43.68	26
149	Zimbabwe	43.41	27
150	Cambodia	43.23	19
151	Solomon Islands	43.22	8
152	Iraq	43.20	17
153	Laos	42.94	20
154	Burkina Faso	42.83	28
155	Sierra Leone	42.54	29
156	Gambia	42.42	30
157	Republic of Congo	42.39	31
158	Bosnia and Herzegovina	41.84	29
159	Togo	41.78	32
160	Liberia	41.62	33
161	Cameroon	40.81	34
162	Swaziland	40.32	35
163	Djibouti	40.04	36
164	Papua New Guinea	39.35	21
165	Eritrea	39.34	37
166	Mauritania	39.24	38
167	Benin	38.17	39
168	Afghanistan	37.74	22
169	Pakistan	37.50	23
170	Angola	37.44	40
171	Central African Republic	36.42	41
172	Niger	35.74	42
173	Lesotho	33.78	43
174	Haiti	33.74	12
175	Madagascar	33.73	44
176	Nepal	31.44	24
177	India	30.57	25
178	Dem. Rep. Congo	30.41	45
179	Bangladesh	29.56	26
180	Burundi	27.43	46

Rank, EPI Score, and Regional Standing (REG, shown in color) for 180 countries.

Asia · Latin America · Caribbean · Mid East & N.Africa · E.Europe & Eurasia · Pacific · Europe & N.America · Sub-Saharan Africa

suppliers and the other procurement stakeholders, better to take into account the longer-term environmental impacts associated with the purchase and all associated costs, operating costs and end-of-life of goods and services. The countries must be mindful of international developments in the area of life cycle analysis.

It is encouraged to facilitate Life Cycle Assessment by attempting to quantify the life cycle costs of emissions, resources consumed and environmental and health impacts that are associated with particular goods and services. Methodologies for life cycle costing take into account the four main cost categories of standard financial costings using market-based pricing valuations – *investment, operational, maintenance and end-of-life/disposal expenses, in addition to which they also aim to include in the analysis values in respect of external environmental factors*. An illustration of of life cycle cost vs. initial cost is shown in Figure 5.

Environmental costs may be derived from analyses of environmental impacts, for example, by estimating the quantity of greenhouse gas emissions and the global warming potential associated with emissions of different greenhouse gases, then multiplying it by a value for carbon. In some instances it may be possible to estimate environmental costs in respect of acidification, land use or other quantifiable externality impacts.

Figure 5. Life cycle cost vs. Initial costs
Source: GPP Bhutan. 2018

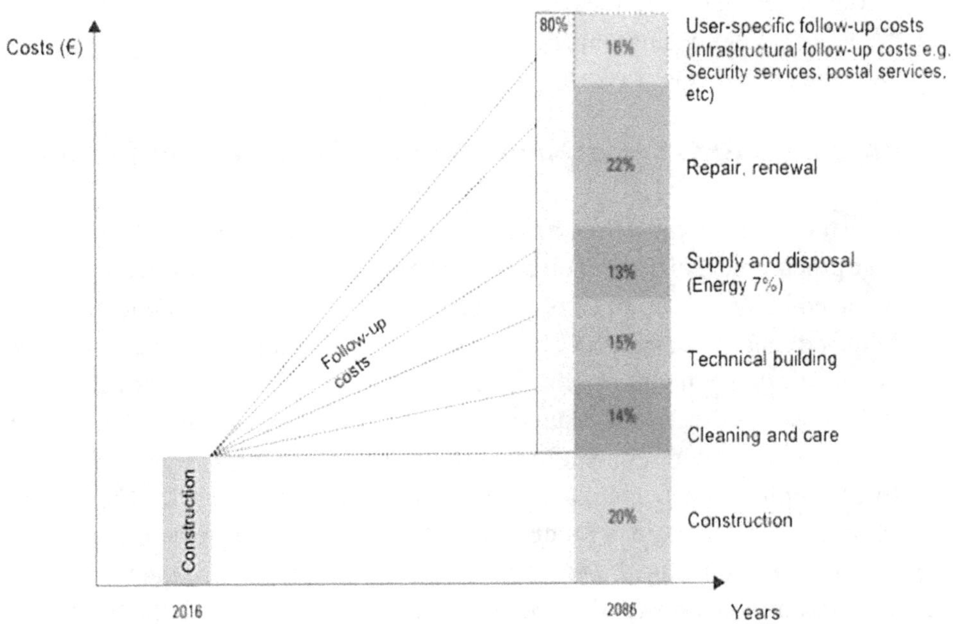

Life cycle analysis is not yet matured, and there is a need for more research into methodologies. For example, impacts on ecosystems are not included in general consideration of environmental life cycle costing "due to lack of data." Nonetheless, it is widely acknowledged that accounting for the cost of carbon, where practicable, is an essential element of such methodologies.

GREEN PUBLIC PROCUREMENT PROCESSES

Objectives of Green Public Procurement

Many countries have set the Green Public Procurement objectives in their policy and legislation, and those are most common. The purpose of such policy intervention is to have a positive impact on the environment, social behavior, and practices, and promote the innovative culture in public procurement. The set of such common objectives in GPP are:

- Adherence to the principle of fit for purpose and cost-effectiveness
- Acquisition of goods that are durable, repairable, have reduced hazardous substances and are re-usable
- Minimizing the use of fossil fuels
- Cost saving on energy consumptions
- Reducing carbon emissions
- Keeping a healthy environment
- Manage and recycle waste

Incorporating Green Provisions Into the Procurement Process

All public procurement must follow the objectives laid out in the procurement plan. While the procurement officer involved in the procurement must consider mitigating risks involved in the procurement of products with green criteria, a value of money also should be equally considered. The readiness of market supplying green products is essential before making the green criteria mandatory. Initially, the green criteria should be chosen as the desired requirements, and gradually make those mandatory when the market is more competitive with the green products, and an adequate number of suppliers are available in the market to bid for the contract. For making the market ready for the green procurement, the government at national level and procuring entities at the local level should develop comprehensive knowledge and capacity building programs, and encourage the suppliers to learn to build their green portfolio. Demand from the government side and frequent opportunities make the suppliers get ready to supply and compete.

In general, green criteria can be used with any types of contracts, but the better effectiveness of its use is observed in the following types of contracts:

- **Framework Agreements:** Framework agreements are mostly used for the common-use items, and the volume of procurements in such procurements are usually high if aggregated across the procuring entities or procured at the national level. Framework agreements can be seen as the tools for cost saving because of its economies of scale. Green criteria can be in the products which are most frequently used at the daily basis, and may directly impact on the daily life of the people.
- **Public-Private Partnership (PPP):** Public-Private Partnerships are usually large and complex contracts which usually involves goods, works, services as well as the share of responsibilities between the government and private parties. The green responsibility of both sides can be effectively incorporated at different stages of the procurement and the implementation processes.

Procuring entities must specify the following information adequately for a clear understanding of the requirements by the bidders:

- A clear description of the item to be procured;
- the technical specifications that can be complied with by the available market;
- the eligibility criteria to participate in the bidding process;
- the contract award criteria with evaluation factors; and
- the provisions for the contract execution phase.

The first step to incorporate GPP in the procurement is by specifying an indication of green procurement in the definition of the items to be procured (EC, 2008). However, GPP requirements (provided these are by the law) may be specified for other criteria as well. For example, where relevant to the items of procurement in the contract, bidder selection criteria, and bidders performance requirements in compliance with the green requirements during contract execution.

About the Award Criteria, contracts may be awarded by either (i) the lowest price or (ii) the most economically advantageous tender (MEAT). Where contracts are to be awarded by MEAT, environmental criteria included in the Award Criteria must be:

- Explicitly specified in the bidding document
- Related to the item of procurement
- Objectively quantifiable and verifiable

- Partial compliance of the requirement should be factored at the early phase of GPP implementation

CONCLUSION

Green Public Procurement is still at its infancy, and many countries are just recently trying to define their policy, strategies and incorporating in the legislation. Only the way to save and enhance our lives in the Earth is by protecting our environment through responsible use and optimal re-use of available resources discouraging production and use of harmful to our health and our environment.

Procurement being the high spending sector of the government should make a meaningful impact on the use of goods, works and services considering environmental, social and financial aspects. Right policy, strategies on GPP, and the seamless incorporation of those policies and strategies in the cycle of procurement processes help ensure accountability of the government to the society and the whole humanity at large.

Policy support in the form of development of a Green Public Procurement Plan or issuance of an official notice to give a concrete signal about the governments' commitment in this regard would be needed. At the beginning any requirement to adhere to green procurement practices should be completely voluntary and used selectively only for projects that are heavily reliant on international procurement and market ready items.

The governments should support the development of a domestic market for green items through measures like easing access to green finance for potential suppliers of green items and providing them guarantee of buying their produced items, fulfilling the government's green technical requirements. Training of procurement staff and suppliers would need to be established as a core function of the Procurement Authorities and continued on a regular basis.

REFERENCES

Department of Expenditure and Reform. (2014). *Green Tenders: An Action Plan on Green Public Procurement.* Available at http://etenders.gov.ie/Media/Default/SiteContent/LegislationGuides/13.%20Green%20Tenders%20-%20An%20Action%20Plan%20on%20Green%20Public%20Procurement.pdf

EC. (2008). *A toolkit developed for the European Commission Local Governments for Sustainability. ICLEI.* Available at: http://ec.europa.eu/environment/gpp/pdf/toolkit/module2_legal_framework.pdf

EPI. (2018). *Environmental Performance Index Report.* Yale Center for Environmental Law & Policy, Yale University Center for International Earth Science Information Network, Columbia University. Available at: https://epi.envirocenter.yale.edu/downloads/epi2018policymakerssummaryv01.pdf

European Commission. (2016). *EU GPP Criteria. Environment.* Author.

GPP Bhutan. (2018). *Green Procurement in Bhutan.* Available at http://gppbhutan.bt/

HSE. (2018). *Implementation of Green Procurement for HSE Procurement/Supplies Officers.* Available at https://www.hse.ie/eng/services/news/newsfeatures/healthsustainabilityoffice/greenprocurement/gpp%20factsheet.pdf

Indiana University. (2018). *Sustainable Purchasing.* Available at http://www.indiana.edu/~purchase/sustainability/sustainability.php

Made-by. (2018). *EU Ecolabel for Textiles. Made-By.org.* Available at: http://www.made-by.org/consultancy/standards/eu-ecolabel-for-textiles/

PPRC-Kosovo. (2014). *National Public Procurement Strategy.* Available at https://krpp.rks-gov.net/krpp/PageFiles/File/draftLS/2014/00-National%20Public%20Procurement%20Strategy-v3_ENG.docx

PwC. (2009). *Collection of Statistical Information on Green Public Procurement in the EU.* Available at http://ec.europa.eu/environment/gpp/pdf/statistical_information.pdf

Shakya, R. K. (2016). *Green Public Procurement Strategy for the East African Community (EAC).* Arusha, Tanzania: EAC.

Siemens, R. (2003). A Review and critical evaluation of selected greener public purchasing programmes and policies. In *The Environmental Performance of Public Procurement. Issues of Policy Coherence* (pp. 51–95). Available at http://www.keepeek.com/Digital-Asset-Management/oecd/environment/the-environmental-performance-of-public-procurement_9789264101562-en#page50

Šličiuvienė, D. (2017). Public procurement in Lithuania: (dis)balance between profitability and environmental protection. In K. V. Thai (Ed.), *Global Public Procurement Theories and Practices, Public Administration, Governance and Globalization* (pp. 17–38). New York, NY: Springer. doi:10.1007/978-3-319-49280-3_2

The Public Contracts Regulations. (2015). Available at https://link.springer.com/chapter/10.1007/978-3-319-49280-3_2/fulltext.html

UN Environment DTIE. (2012). *The Impacts of Sustainable Procurement: Eight illustrative case studies*. Retrieved from http://www.unep.fr/scp/procurement/docsres/projectinfo/studyonimpactsofspp.pdf

UNEP. (2011). *Buying for the better world: A guide on sustainable procurement for the UN system*. Paris, France: UNEP.

UNESCAPE. (2012). *Low Carbon Green Growth Roadmap for Asia and the Pacific*. Available at http://www.unescap.org/sites/default/files/33.%20FS-Green-Public-Procurement.pdf

Chapter 2
Implementation of Green Considerations in Public Procurement:
A Means to Promote Sustainable Development

Andrea Appolloni
University of Rome "Tor Vergata," Italy

Maria Antonietta Coppola
Independent Researcher, Italy

Gustavo Piga
University of Rome "Tor Vergata," Italy

ABSTRACT

Green considerations can be applied during all phases of the public procurement process: from the pre-award to the award and post-award phase. They can be included in technical specifications, award criteria, and contract performance clauses. Technical specifications provide a detailed description to the market of the good, work, or service to be procured. They constitute the basis for drafting green award criteria, which allow contracting entities to evaluate the received bids and award the contract. Contract performance clauses can also be used by public contracting entities to introduce environmental considerations in the procurement process. These clauses are based on the capacity of the winning bidder to perform the negotiated environmental criteria. Notwithstanding the importance of GPP, green considerations are seldom applied in public procurement. Reasons can be found in the lack of appropriate regulations at national and international level, or in the actual lack of training of the procurement workforce.

DOI: 10.4018/978-1-5225-7083-7.ch002

INTRODUCTION

Procurement can be defined as the activity of purchasing goods, works and services in a transparent, impartial, non-discriminatory and competitive manner in order to achieve both the quality of outputs and value for money. Going more into details, it can be stated that public procurement has a strategic function for public spending, as it accounts for a large amount of countries' gross domestic product (GDP). As a matter of fact, there are several sectors in which the use of public funds to procure goods, works and services is prominent, these are: health, education, social protection, waste management, energy and transport. These sectors are tightly linked to the concept of sustainable development, which can be defined as multidimensional. Indeed, sustainable development encompasses economic, social and environmental aspects that must be considered as part of the same whole. Given its broad implication, sustainable development should be pursued both at national and international level, as no national boundaries apply to it.

This chapter focuses on the use of the environmental aspect of public procurement to achieve sustainable development. To this aim, the chapter explores green public procurement (GPP), a voluntary instrument that can be used by public authorities to award contracts on the basis of green criteria. More specifically, it can be defined as a process whereby contracting entities seek to procure goods, works and services that have a reduced environmental impact throughout the entire life-cycle when compared to goods, works and services that would otherwise be procured.

GPP can be applied during all phases of the public procurement process, from the pre-award to the award and post-award phase. In fact, green criteria can be included in technical specifications, award criteria and in contract performance clauses. Technical specifications provide a detailed description to the market of the good, work or service to be procured and can be considered a starting point to introduce sustainable considerations in the procurement process. In fact, they constitute the basis for drafting green award criteria, which allow contracting entities to evaluate the received bids and award the contract. Contract performance clauses can also be used by public contracting entities to introduce environmental considerations in the procurement process. These clauses are based on the capacity of the winning bidder to perform the negotiated environmental criteria.

Notwithstanding the importance of public procurement in the pursuit to sustainable development, sustainable considerations are seldom applied in the procurement process. One of the reasons can be found in the lack of appropriate regulation of environmental aspects in the public procurement national and international legal frameworks, which not always provide for clear guidelines to contracting entities on how to apply them. Another reason is the actual lack of training and information of the procurement workforce which is, therefore, less keen in introducing environmental criteria in the procurement process.

The objectives of the chapter are therefore to provide the reader with a general understanding on the application of GPP during the whole procurement process and to suggest the implementation of appropriate national and international regulations coupled with adequate training programmes as adequate means to promote the use of GPP.

GREEN PUBLIC PROCUREMENT AND SUSTAINABLE DEVELOPMENT: DEFINITIONS, LINKS AND INTERNATIONAL LEGAL BACKGROUND

Procurement can be defined as the activity of purchasing goods, works and services in a transparent, impartial, non-discriminatory and competitive manner in order to achieve both the quality of outputs and value for money. This definition entails the main principles of every procurement process, which also constitute the core part of the international legal framework for public procurement.

The principles mentioned in the definition reported above should be promoted in each phase of the procurement cycle, from the pre-award phase, to the award and post-award phase. For what concerns transparency, it should be fostered in order to allow access to procurement information without underestimating the importance of protecting trade secrets and proprietary information. Two other important principles are non-discrimination and equal treatment, which forbid the discrimination of parties on the basis of their nationality or affiliation to any group of firms, and encourage contracting entities to treat equally every participant to the procurement process. A fourth principle is competition, according to which all tenderers must have an equal chance to participate in a procurement process where competition must not be artificially narrowed nor distorted. Transparency, non-discrimination, equal treatment and competition are the underlying principles to the international legal framework for public procurement, which can be considered as the basis for every primary and secondary law aiming at regulating procurement at national level.

But why is procurement relevant for sustainable development? Procurement accounts for 15-20 per cent of the global gross domestic product (GDP) thus representing a big portion of the public spending. This number testifies - especially in a period of economic constraints - the strategic relevance of the procurement function and the importance of regulating it in order to provide for sound spending of public money and to influence the market so to introduce sustainable buying criteria.

There are several sectors in which the use of public funds to procure goods, works and services is prominent, these are: health, education, social protection, waste management, energy and transport. These sectors are tightly linked to the concept of sustainable development, as it entails different economic, social and environmental aspects.

To this regard, there are many definitions that illustrate the concept of sustainable development. The landmark definition is certainly the one included in the 1987 Brundtland Report, named "Our Common Future", which states as follows:

Humanity has the ability to make development sustainable to ensure that it meets the needs of the present without compromising the ability of future generations to meet their own needs. (World Commission on Environment and Development, 1987)

This definition implies that meeting the needs of the present generations may jeopardize the fulfilment of the needs of future ones. An example of this antagonistic relationship is the industrial growth that can be in contrast with the need of future generations to preserve natural resources and the ecosystem. This dilemma is a good example to demonstrate the multidimensionality of sustainable development, which encompasses economic, social and environmental needs that must be considered as tightly linked one another. Going more into details, the environmental implications of sustainable development refer to the preservation of the ecosystem, of the biodiversity and natural resources by promoting a better, more sustainable and rational use of such resources.

Given the economic, social and environmental dimensions, it is clear that sustainable development cannot be pursued only at country level rather it should be pursued at international level as no national boundaries apply to it. As a matter of fact, the following paragraphs illustrate the international commitments taken towards the environmental dimension of sustainable development.

The landmark conference in which the international community affirmed its commitment in finding a common strategy was the Earth Summit, held in Rio de Janeiro in June 1992. The summit was organized by the UN Commission on Environmental Development (UNCED) and its main themes were environment and sustainable development. This summit led to the adoption of the United Nations Framework Convention on Climate Change (UNFCCC).

Since UNFCCC adoption, annual conferences of the parties (COP) that ratified the Convention followed up the 1992 Rio Conference in order to implement the Convention and negotiate new commitments. During COP3, on 11 December 1997 in Japan, the Kyoto Protocol was adopted. This is an international agreement linked to the UNFCCC which sets internationally binding emission reduction targets. Its implementation rules were adopted in 2001 during COP7 and are known as the Marrakesh Accords.

More recently, in 2015, it was held COP21, also known as the 2015 Paris Climate Conference. After two weeks of negotiation, on December 2015, the conference led to the adoption of the first universal climate change agreement. The Paris agreement can be considered an historic document in which the 195 Parties to the UNFCCC pledged to curb emissions and strengthen their resiliency to losses and damages derived from climate impacts. In the framework of the conference, each State published its Intended Nationality Determined Contribution (INDC), a commitment to reduce greenhouse gas emissions in a set period of time. Each State had also the possibility to choose the adaptation measures to be undertaken in order to reduce global warming.

To date, the political commitment worldwide may slow down or halt the process recently started at the 2015 Paris Climate Conference, nevertheless it would be desirable for the international community to continue preserving the ability of future generations to meet their needs by undertaking significant commitments towards environmental sustainability.

As stated above, the pledge to promote sustainable development should be common to both developed and developing countries. But how can it be achieved? Concrete measures towards the promotion of sustainable development can be certainly undertaken through a sound use of the procurement function. In this view the term "sustainable procurement" has been created. Procurement is called sustainable when it combines requirements, specifications and criteria that encompass the different dimensions of sustainability. Therefore, it is sustainable when public and private entities use their buying power in a way that it is compatible with social enhancement, the protection of the environment and the economic development.

The introduction of sustainable requirements and criteria in the various phases of the procurement process can certainly lead to more sustainable production and consumption practices. Notwithstanding the importance of the introduction of such criteria and requirements throughout the whole procurement process, it is important to note that sometimes these may be considered as a bundle both by contracting entities and bidders. In addition, sustainable requirements and criteria ought to be proportionate to the size and complexity of the procurement in order not to distort competition. As a matter of fact, disproportionate and inadequate environmental criteria can have the negative effect to narrow the scope of competition by putting additional economic and/or administrative burdens on bidders. These considerations must be taken into account when trying to implement sustainable development objectives through procurement, especially while reforming the legal framework that regulates the procurement function or while delivering training and building capacity.

The tight link between procurement and sustainable development has been recently confirmed by the United Nations (UN) General Assembly with the adoption of the 2030 Agenda for Sustainable Development on 21 October 2015.

The new development agenda follows up the Millennium Declaration, a landmark declaration approved in the year 2000 by the UN General Assembly during the Millennium Summit. This declaration spelled out the 8 Millennium Development Goals (MDGs), which aimed to eradicate poverty and hunger and promote equality throughout the world by the year 2015. Fifteen years later, in 2015, because many of the MDGs were partially met, the international community has approved a new development agenda in order to expand the set goals and to outline a more comprehensive international action plan for sustainable development.

The new development goals are 17 with 169 associated targets which are indivisible from the set action plan. These goals came into force starting from 2016, and establish a clear and comprehensive agenda to be implemented in the following 15 years.

What is important to note on the procurement side, is that by approving Resolution A/RES/70/1 the international community has officially recognized the link between sustainable development and public procurement, with a special focus on national policies and priorities. This tight link is clearly stated in target 12.7, in which it is affirmed that the sustainable consumption and production patterns can be achieved by promoting sustainable public procurement practices.

Goal 12: Ensure sustainable consumption and productions patterns.

Target 12.7: Promote public procurement practices that are sustainable, in accordance with national policies and priorities. (United Nations General Assembly, 2015)

Notwithstanding the novelty of this development agenda, it can certainly be affirmed that sound procurement is indispensable both in theory and in practice to achieve all SDGs.

As it has been recognized in the 2030 Agenda for Sustainable Development, procurement is a tool that should be utilized to foster sustainable development. Through procurement, national and international agencies and institutions have the possibility to positively influence national and international markets, thus enhancing their development, by introducing green ad social considerations such as the reduction of CO2 emissions or the promotion of labour rights. For this reason, harmonized strategies and regulations to implement the aforementioned Agenda through the procurement function should be adopted.

A way to foster such strategies and regulations is the endorsement of a common international legal framework that regulates procurement and that is reflected in national primary and secondary laws. The agreement that establishes common policies to harmonize the main aspects of public procurement among different countries is the Agreement on Government Procurement (GPA). This is a multilateral agreement adopted within the framework of the World Trade Organization which has the aim to ensure open, transparent and fair competition in public procurement (World Trade Organization, 2014). The text of the GPA has recently been revised and adapted to the most recent international practice. Its latest version entered into force on 6 April 2014. In addition to the GPA, an example of regulations adopted at regional level is the public procurement legislative package of the European Union. More specifically, Directive 2014/24/EU reflects the obligations included in the GPA and regulates the main aspects of all the phases of the procurement process also taking into account sustainable and environmental considerations.

In addition to the aforementioned regulations, it is worth mentioning two non-binding documents endorsed by the international community that outline public procurement principles and procedures. UNCITRAL Model Law on Public Procurement is a document drafted by the UN Commission on International Trade Law (UNCITRAL) that is designed to assist States in reforming their public procurement systems. The text of the model law has been amended in 2011 to replace the 1994 UNCITRAL Model Law on Procurement of Goods, Construction and Services. Another important document is the 2015 OECD Recommendation on Public Procurement, adopted by the Organisation for Economic Co-operation and Development (OECD), an international economic organisation whose mission is to promote economic and social well-being (Organisation for Economic Co-operation and Development, 2015). This document promotes the use of a strategic and holistic approach for public procurement in order to support the efficient allocation of public resources and mitigate the risk of inefficiency and corruption.

Given the abovementioned legal framework, in order for the procurement function to foster sustainable development, national and international reforms should be coordinated. To do so, they should address the whole procurement cycle and be consistent with the main legal documents adopted at international and regional level. Such reforms should be further supported by the allocation of funds for the provision of adequate training to build capacity of human resources.

IMPLEMENTATION OF GREEN PUBLIC PROUREMENT IN THE PROCUREMENT PROCESS

The adjective "green" referred to products and services indicates that such products and services have the following characteristics when compared to similar ones: require the use of fewer resources; have a lower impact on the environment; are designed to last longer. These characteristics imply that green products have a reduced impact on human health but require a greater upfront expense that can lead to the saving of money in the overall life-cycle of the products.

GPP policies can be applied by any organization, regardless to the size, and can vary from the procurement of green supplies, such as recycled paper, to the procurement of complex works and services based on specific environmentally friendly requirements for suppliers and contractors. In order to be implemented, green purchasing policies should be assessed taking into account the objectives of the specific organization and the green criteria against which the outcome of such policies should be measured. The adoption and implementation of green procurement policies can bring organizations to reduce costs, waste and increase efficiency. Moreover, purchasing policies have the power to influence the market by stimulating research and development (R&D), the production of green products, the fulfilment of green requirements thus leading to better prices and increased availability of green products.

GPP is playing an increasingly important role in the demand and production of products and services that are environmentally friendly as the overall impacts of government operations can exceed the ones of individual industry sector. As a matter of fact, public agencies are the major employers, providers of services and consumers of resources and therefore the inclusion of environmental aspects in their practices is of outmost importance to achieve national and international sustainability goals.

Going more into details, GPP is a voluntary instrument that can be used by public authorities to award contracts on the basis of green criteria. The European Union defines it as a process whereby contracting entities seek to procure goods, works and services that have a reduced environmental impact throughout the entire life-cycle when compared to goods, works and services that would otherwise be procured.

There are several benefits that can derive from GPP, such as the provision of innovative solutions. In fact, allowing for the inclusion of green criteria and requirements in the tender process can definitely be an incentive for the industry to invest in the creation of innovative products and to introduce them to the market. Also, GPP can increase financial savings of public funds, especially when the full life-cycle cost is taken into account and not just the purchasing price. Some examples can be the reduction of electricity bills when using green energy or the reduction of the costs of the disposal of hazardous waste. Furthermore, in the long run GPP can bring significant advantages not just to the contracting entity but to the whole society. In fact, in addition to reducing the life-cycle cost of the goods and services procured, it also reduces the future costs of adjustments that will derive from pollution and climate change.

Unfortunately, many obstacles lie ahead on the road to a sound implementation of GPP. Among them, there are the initial costs that economic operators have to bear in order to invest in environmentally friendly products and services, such as investments in R&D. These costs can cause the reduction of competition as not many economic operators may be able to make such investments to match green requirements. Therefore, in order to promote GPP and preserve the basic principles of procurement during the whole procurement process, specific mutually reinforcing measures should be undertaken. One of them is improving the environmental performance of products and services during their whole life-cycle, which should be coupled with the promotion on the market of a greater demand for greener products and technologies. Such measures should be supported by well-tailored communications policies that can help inform contracting entities, and consumers at large, on how to make greener choices.

Given the benefits and obstacles to GPP, it is very important to note that green criteria can be introduced in every phase of the procurement cycle, from the pre award to the award and post-award phase. As a matter of fact, they can be included in specifications, selection and award criteria, or in the contract performance clauses. But how green criteria can be set? International, regional, national and local authorities have developed sets of green criteria. One of them is the European Union (EU), which has developed a set of green criteria for a number of product and service groups, among others: copying and graphic paper; office IT equipment; constructions; transport; furniture; street lighting and traffic signs. These criteria can be directly incorporated in the tender documents and include information related to the possible methods to be used in order to verify green requirements. Green criteria are based on information collected from the main stakeholders, the civil society and public authorities, on evidence-based data and eco-labels.

Green Public Procurement in the Pre-Award Phase

As mentioned in the previous paragraphs, public agencies are the major employers, providers of services and consumers of resources and therefore the inclusion of environmental aspects in their practices is of outmost importance to achieve national and international sustainability goals. Nevertheless, contracting entities have freedom to decide whether to include environmental considerations in their procurement. This decision should take into account several factors, such as the subject matter of the contract, market analysis, consultations with relevant stakeholders. In the beginning of the procurement cycle, contracting authorities define the subject matter of the contract usually providing a description of the goods, works or services to be procured. In order to do so, it is important to define the actual needs by conducting a detailed market analysis which illustrates market dynamics. This analysis is essential for contracting entities to make the right procurement choices such as dividing the procurement asymmetrically with respect to the composition of the market. In fact, in case of an opposite decision, the contracting entity may cause the distortion of the level playing-field for the competition. In addition to market analysis, proper consultations with internal users and end users can be organized by procuring entities to assess the actual needs and their environmental impacts.

Both market analysis and consultations with relevant stakeholders are essential to understand the environmental impacts of the goods, works and services to be procured. The European Union has outlined a set of potential environmental impacts for each type of contract. For what concerns supply contracts, the set of environmental impacts includes the materials (e.g. renewable materials) and the natural resources (e.g. fuel and water consumed for the production of the good to be procured) used to produce the goods, the durability of the goods produced and the opportunities to recycle them. Also, consideration is given to the way goods are packed and transported. With regard to service contracts, different aspects are taken into account: the technical expertise and qualification of the staff; the products used to perform the service to provide; the amount of energy and water consumed to perform the service together with the generated waste; the way in which procedures are managed so to minimize the environmental impact of the service provided. In addition to the aforementioned considerations made for supply and service contracts, works contracts have additional environmental impacts that should be taken into account (e.g. different impacts related to the planning of traffic or to the use of land) depending on the type of work to be performed.

Once the subject matter of the contract and its environmental impacts are defined, contracting entities need to outline measurable technical specifications that are proportionate to the size and complexity of the procurement. Technical specifications have the function to provide a detailed description to the market of the good, work, or service to be procured, thus allowing interested economic operators to draft appropriate bids. In addition, they provide measurable requirements against which such bids are evaluated. Therefore, they can be considered minimum compliance criteria that can lead bidding firms to draw unsuitable offers in case they are not correct or specific enough. Technical specifications can be defined in relation to technical standards or performance requirements.

Contracting entities may refer to environmental standards in technical specifications, which are useful information sources that shall be appropriate and proportionate to the subject matter of the contract and shall by no mean hinder competition. To this regard, when technical specifications refer to specific standards these shall be followed by the words "or equivalent". Ecolabels are an example of environmental technical standards and certify that a product or a service meets specific environmental requirements. These labels measure the environmental impact of a product based on scientific evidence and help contracting entities to set – and subsequently evaluate – specific environmental technical specifications. It is important to note that in order to preserve competition, ecolabel criteria shall not be mandatory and therefore procurers should indicate that the criteria underpinning the label shall be met, but should not require bidders to hold a specific ecolabel. It is also important to specify that, ecolabels refer to the characteristic of the products and services and to their production processes whilst they are not related to the general management of the bidding firm.

In addition to technical standards, environmental considerations can also be included in performance-based specifications. A performance-based specification usually describes the desired results and outputs and indicates how they will be measured. This type of specification does not include the inputs to be provided or the methods to be used by economic operators. The assessment and comparison of performance-based specifications should be done in a fair and transparent way and in order to do so, contracting entities may ask tenders to explicitly indicate in their bidding documents the inputs they would provide in order to achieve the set outputs.

Environmental technical specifications are an important tool to achieve sustainable development through the procurement function. Nevertheless, contracting entities may be less keen in using green requirements because of the difficulty in assessing during the award phase the compliance of the bids received without the support of technical evaluators. To overcome these obstacles, environmental product declarations can be used as a way to verify compliance with environmental technical specifications. ISO 14000:2004 and EMAS, among others, are voluntary environmental management instruments that are widely used as environmental product declarations. They are international standards used to certify the environmental management systems (EMS) of the environmental programs of any organization. Their aim is to help assess, manage and continuously improve organizations' environmental performance. As for ecolabels, also in the case of EMS it is important to note that in procurement it is not allowed to require bidding economic operators to hold a specific environmental product declaration without allowing for an equivalent certification.

Green Public Procurement in the Award Phase

During the award phase the contracting entity mainly evaluates and compares the quality and cost of the received tenders. This phase is also characterized, where in place, by post-award debriefings, eventual modification and publication of the contract, and claims related to the selection and contract award process. The public procurement systems of different countries may differ one another, therefore some may require – for example – for the publication of the bids received but others may not. As for the previous phase, the use of GPP criteria during the award phase can help contracting entities to promote environmental policies that help countries reach their sustainability goals.

The award of a public contract is done according to the award criteria which shall be predetermined, set in advance and linked to the subject matter of the contract. It is important to note that the ability to meet each award criteria is not a prerequisite for being a candidate for the contract, while this cannot be affirmed for the selection and exclusion criteria. Award criteria enable the contracting entities to evaluate and compare the different bids received based on price solely or on price and quality. When the winning bid is evaluated based solely on price, the lowest bid is awarded the contract. On the contrary, when the award criteria also takes quality into account, the contract is awarded to the most economically advantageous tender (MEAT) based on specific scoring formulas. In some cases, it may happen that contracting entities receive abnormally low bids, and therefore it is important to have a proper procurement system in place which helps identify, and whenever necessary reject, bids that have an abnormal low price.

Award criteria under MEAT may include objective and verifiable environmental criteria which may not be solely based on purely economic factors. In this case the contracting entities weight each bid based on the different award criteria published in the procurement notice and subsequently compare the relative advantages of the different bids.

For what concerns environmental award criteria, as stated by the European Commission in its handbook on green public procurement:

A number of considerations should be taken into account when assessing whether an environmental characteristic should be a minimum requirement (specifications) or a preference (award criterion). (European Commission, 2016)

The former is related to the ability of the contractor rather the latter is related to its offer.

The inclusion of green award criteria leads contracting entities to value more sustainable tenders over less sustainable ones as the bidder who can deliver the good, work or service to be procured in a more sustainable way – which not necessarily is the less expensive one – gets the contract awarded. Furthermore, green award criteria can be used by public authorities to promote horizontal procurement objectives that are not necessarily connected with the function of the item to be procured. The use of horizontal policies as award criteria can encourage the market to innovate and provide more sustainable solutions which go beyond the established standards.

However, the use of environmental award criteria can cause uncertainty. It has been argued that environmental externalities do not bring a direct economic advantage to the contracting entity, and therefore they can be in conflict with the MEAT principle. The Court of Justice of the European Union (CJEU) has contributed to clarify the legitimacy of the use of sustainable award criteria by ruling on milestone cases such as the Concordia Bus and Wienstrom (Court of Justice of the European Union, 2002, 2003). In the Concordia Bus case the CJEU clarified that – for the purpose of including horizontal objectives into public procurement policies – costs that are generally considered as environmental externalities can be included in the factors to be considered in order to determine the MEAT as long as they are linked to the subject matter of the procurement. In the Wienstrom case, the CJEU further clarified that award criteria based on horizontal policies can only be considered unlawful in case they violate the general procurement principles of proportionality, transparency and equal treatment. On the contrary, the inability of the contracting entity to prove that the set award criteria do not serve to achieve horizontal objectives is not relevant to consider the set criteria unlawful. Therefore, the contracting authority has discretion in determining the award criteria given that its discretion is used to identify the MEAT without any unjustified distortion. In turn, the contracting entity needs to be able to evaluate the received tenders in a transparent and objective manner based on the proof and information provided in the tenders received.

But how is it possible to apply environmental award criteria? During the award stage the selection committee needs to evaluate the offers received and to score them based on the public procurement principles of competition, equal treatment and non-discrimination. To this regard, it is very important that award criteria do not duplicate any assessment which has taken place during the qualification stage. A mean to introduce environmental award criteria is to set a minimum level of performance in technical specifications and, subsequently, allocate additional points to bidders who offer a better performance. Similarly, it is also possible to set award criteria based on the environmental criteria on which eco-labels are based. These criteria may help to draft and assess environmental award criteria. For example, contracting entities may refer to specific environmental considerations underpinning specific eco-labels in the technical specifications, and subsequently allocate extra points to products or services that meet more of the relevant environmental criteria included in the eco-labels. In other cases, environmental management systems can also be considered as evidence when assessing environmental award criteria since during this stage contracting entities need to take into consideration how the contract will be performed. As for technical specifications, eco-labels and EMS cannot be mandatory, therefore the contracting entity needs to refer to the principles underpinning these certifications rather than insist on a particular certification or label.

The most influential factor during the award stage is the cost of a tender. Contracting entities need to be able to define the cost of each tender, which corresponds to the price of the good, work or service offered in the received bids. With regard to the role of the procurement function for the promotion of sustainable development, it is important to note that this cost does not only refer to the purchase price of what is being procured. The actual cost of each tender corresponds to the sum of the costs that the contracting entity will bear during the whole process of procuring, owning and disposing. This is intended as the life-cycle costing (LCC) and takes into considerations the costs of the good, work or service to be procured during its whole life-cycle. The costs to be considered are: the purchasing price and the associated costs; operating costs; end-of-life costs. LCC also takes into account the costs of externalities such as greenhouse gas emissions.

The main potential cost savings with regard to GPP over the life-cycle of a good, work or service to be procured are: savings on the use of energy and natural resources; savings on disposal costs; savings on maintenance and replacement. These costs are directly borne by the contracting entity. In addition, other types of costs exist, these are the so called indirect costs, which are borne by society at large because of the environmental externalities of what is being procured. In order to be assessed during the award phase, indirect costs shall be expressed in monetary terms, they shall also be linked to the subject-matter of the contract and abide to the general principles for the award criteria. LCC that takes also the cost of externalities into account is referred to as whole-life costing (WLC).

Green Public Procurement in the Post-Award Phase

The post-award phase, also known as contract-performance phase, is the phase where contracting entities have more discretion in introducing sustainable considerations. In the post-award phase green and social considerations are usually included in the contract performance clauses. These specific conditions of the contract cannot be considered upfront during the previous phases and therefore it is not possible to include them in the technical specifications or contract award criteria. As a matter of fact, specific conditions of a contract are related to contract deliverables, together with the capability of an economic operator to comply with such requirements, and cannot be set before the award of a contract. There are two important aspects to take into account during the contract-performance phase: the difference among technical specifications, award criteria and the contract performance clauses; the risk of contractor's non-compliance with the contract performance clauses.

In the previous paragraphs the difference between technical specifications and award criteria has already been outlined. But how do they differ from contract performance clauses? The contract performance clauses are included in the contract that has to be signed between the contracting entity and the winning bidder and specify the way in which a contract should be carried out. These clauses shall comply with national and international laws in place. Furthermore, in many procurement systems, they shall be included in the contract notice so as to inform bidders about the conditions of the contract and allow them to reflect these specific conditions in their bids. It is important to note that, during the selection process, contracting entities cannot require economic operators to comply with the contract performance clauses and therefore exclude bidders on the basis of their inability to comply (as for selection or award criteria). Such clauses shall be linked to the performance of the contract and shall be set in a way that they do not favor any bidder.

With regard to the inclusion of sustainable considerations in the procurement process, different types of contracts may include different types of contract performance clauses. As for contracts related to the procurement of goods, the contracting authority may require the winning firm to deliver the goods to be procured outside of peak traffic time, it may also require the firm to provide information related to the greenhouse gas emissions produced to transport the aforementioned goods. In addition, the contracting entity may ask the contractor to reduce or recycle the packaging material already used. For what concerns the procurement of works and services, the contract may include clauses related to the minimization of the waste produced, the efficient consumption of natural resources or chemical products such as detergents. Moreover, the contract clauses may include the application of specific environmental management measures, in accordance – where appropriate – to third party certified schemes such as ISO 14000:2004.

Monitoring is an essential element of the contract-performance phase as it allows the contracting entity to check the compliance of contractors to the specific conditions of the contract. To favor the respect and application of the contract performance clauses that introduce sustainable considerations, the contracting entity may request the contractor to provide evidence of compliance. The contracting authority may also carry out random checks as well as contract a third party to monitor compliance in its place. To conclude, enforcing measures can be taken to encourage contractors to respect the specific conditions of the contract, such as the provision of penalties in case of non-compliance and bonuses in case of good performance.

REGULATIONS AND TRAINING: MEANS TO ENCOURAGE THE IMPLEMENTATION OF GREEN PUBLIC PROCUREMENT

As stated in the previous paragraphs, procurement is an essential tool to promote sustainable development especially when it comes to its environmental dimension. Nevertheless, the lack of appropriate regulations on GPP may not provide for clear guidelines to contracting entities on how to apply environmental criteria and therefore may not encourage the use of green technical specification, award criteria or contract performance clauses. This may especially be true in the case of complex procurement. In this light, the recent international legal framework that regulates public procurement includes references to GPP to foster its implementation.

One example is the revised GPA. In its text the Parties agree to draft and use technical specifications and evaluation criteria that have the scope to preserve natural resources and protect the environment. Such references constitute a common commitment for the implementation of GPP policies to be implemented at national level both by developed and developing countries. However, a lot more needs to be done to favour coordination and the application of the same environmental principles among different countries.

Apart from the policy level, the implementation of GPP can also be encouraged by building capacity of the procurement workforce. In this field, capacity building has the main functions to raise awareness on the social and economic benefits of buying green as well as to improve technical knowledge on the use of green considerations in the procurement process.

To this regard, it can be argued that the lack of training and information on GPP may not encourage the application of green considerations in the different phases of the procurement process. For example, the lack of knowledge on buying green may lead procurement officials not to take such aspect into considerations when formulating the needs and/or doing market research. Another aspect is that procurement officials may not be keen in introducing green considerations without owning the proper knowledge nor competence on the matter. The lack of knowledge and training may also lead to the inadequate use of green criteria in the procurement process thus leading – for example – to the use of environmental criteria that narrow or distort the scope of competition, or to the imposition of additional economic burdens on bidders which may not be proportionate do the size or scope of the procurement.

Therefore, adequate training to build capacity of the procurement workforce is essential to promote the use of GPP. To do so, cooperation among institutions, Universities, training centres should be promoted to implement adequate programs and/or symposia for knowledge sharing among practitioners and academia. Such initiatives could establish fruitful networks and debates as they blend the theoretical side of public procurement together with its practice.

GREEN PUBLIC PROCUREMENT: AN INSIGHT ON LIFE CYCLE COSTING

Among the most controversial aspects of GPP there is the evaluation of the potential cost savings that can be obtained by the procurement of green goods, services and works. Such savings are deemed a significant incentive for the use of GPP by contracting entities but can be of difficult calculation. More specifically, it can be hard to identify and calculate direct and indirect costs of the environmental aspects introduced in the procurement process without proper knowledge on the matter. To support practitioners in such an exercise, LLC and WLC are essential instruments to be used.

As already mentioned in the previous paragraphs, the main potential cost savings with regard to GPP over the life-cycle of a good, work or service to be procured are: savings on the use of energy and natural resources; savings on disposal costs; savings on maintenance and replacement. These costs are directly borne by the contracting entity and are related to the acquisition, use, maintenance and end of life of the good, service or work to be procured. There are also other types of costs that are linked to the subject matter of the procurement. These are indirect costs which are borne by society at large due to the environmental externalities of the good, service or work to be procured. In both cases, the assessment of the procurement costs is not an easy exercise and requires competences and sector-specific knowledge.

The importance of assessing life cycle costs has been affirmed by relevant international and regional institutions, among them there is the European Union. In its Directive 24/2014 the EU allows for the use of the life cycle costing approach when determining the most advantageous tender and encourages the creation of common methodologies for LCC. To this regard, the explicit inclusion in binding national and international regulations of the possibility to assess the costs of the procurement taking into account the costs linked to the life cycle of the good, work or service to be procured can definitely encourage the implementation of GPP.

Nevertheless, as the calculation of the savings green procurement can bring both to contracting authorities and society at large is not an easy exercise, common sector-specific guidelines and methodologies for calculating life cycle costs should be further defined with the aim to be shared among academia and practitioners.

CONCLUSION

As it has been highlighted in the 2030 Agenda for Sustainable Development, public procurement can certainly be considered an essential instrument to promote sustainable development. More specifically, in order for the procurement function to foster sustainable development, specific policy instruments can be used. Among them there is GPP, a voluntary instrument used by public authorities to include environmental considerations in all phases of the procurement process: pre-award, award and post-award phase.

Sustainable considerations can be included in technical specifications, award criteria and in contract performance clauses. Technical specifications provide a detailed description to the market of the good, work or service to be procured and can be considered as a starting point to introduce sustainable considerations in the procurement process. In fact, they constitute the basis for drafting green award criteria, which allow contracting entities to evaluate the received bids and award the contract. Contract performance clauses can also be used by public contracting entities to introduce environmental considerations in the procurement process. These clauses are based on the capacity of the winning bidder to perform the negotiated sustainable criteria.

The adoption of green consideration in the procurement process can bring several benefits to contracting entities, whose objective is to obtain better value for the money spent. Also society can benefit from the use of GPP, thanks to a more sound use of taxpayers' money and to the preservation of natural resources and the ecosystem. As a matter of fact, green criteria aim at reducing the costs of future adjustments that will derive from pollution and climate change.

In addition, GPP can be used to stimulate research and development, by encouraging public authorities to look for more efficient solutions to fulfill their needs and, at the same time, encouraging the market to provide innovative solutions based on green requirements.

With regard to the economic benefits the GPP can bring, the use of green criteria can certainly increase financial savings of public funds especially when the reduction of life-cycle costs is taken into account and not just the reduction of purchasing prices. LCC is very difficult to assess and common methodologies for calculating it can be very handful when implementing green procurement policies.

Given the complexity of the use of green considerations in the procurement process, it is very important to provide for the capacity building of the public procurement workforce. This can be done by establishing networks and debates between academia and practitioners, by organizing university programs, conferences and encouraging debates.

To conclude, the adoption of green criteria and requirements can affect the achievement of international sustainability goals and stimulate both public and private sector to adopt sustainable practices. These concepts should be a key to the reading and interpretation of the international legal background that regulates public procurement. In this light, the inclusion of GPP in the horizontal policies of the countries that commit to the 2030 Agenda as well as in international regulations should be a goal to be pursued both by developed and developing countries in order to achieve a more sustainable growth and guarantee a better future for future generations to come.

REFERENCES

Court of Justice of the European Union (CJEU). (2003). Case No. C-448/01. EVN AG and Wienstrom GmbH v Austria (Wienstrom). ECR I-4527.

Court of Justice of the European Union (CJEU). (2002). Case No. C-513/99. Concordia Bus Finland v Helsingin Kaupunki (Concordia Bus). ECR I-7123.

European Commission (EC). (2016). *Buying Green! - A Handbook on Green Public Procurement*. EC.

Organisation for Economic Co-operation and Development (OECD). (2015). *OECD Recommendation of the Council on Public Procurement*. OECD.

United Nations General Assembly. (2015). *Transforming Our World: the 2030 Agenda for Sustainable Development.* A/RES/70/1. UN.

World Commission on Environment and Development. (1987). *Our Common Future*. Oxford University Press.

World Trade Organization (WTO). (2014). *General Agreement on Public Procurement (GPA)*. WTO.

ADDITIONAL READING

Noguiero, L., & Ramos, T. B. (2014). The Integration of Environmental Practices and Tools in the Portuguese Local Public Administration. *Journal of Cleaner Production*.

Piga, G., & Tatrai, T. (2015). *Public Procurement Policy. Series The Economics of Legal Relationships*. Routledge. doi:10.4324/9781315686356

Piga, G., & Treumer, S. (2013). *The Applied Law and Economics of Public Procurement. Series The Economics of Legal Relationships*. Routledge. doi:10.4324/9780203096314

Rizzi, F., Frey, M., Testa, F., & Appolloni, A. (2014). Environmental value chain in green SME networks: The threat of the Abilene paradox. *Journal of Cleaner Production, 85*, 265–275. doi:10.1016/j.jclepro.2014.09.001

Testa, F., Annunziata, E., Iraldo, F., & Frey, M. (2016). Drawbacks and opportunities of green public procurement: An effective tool for sustainable production. *Journal of Cleaner Production, 112*(Part 3), 1893–1900. doi:10.1016/j.jclepro.2014.09.092

United Nations Environmental Programme (UNEP). (2015). *Measuring and Communicating the benefits of Sustainable Public Procurement (SPP). Baseline Review and Development of a Guidance Framework*. UNEP.

United Nations General Assembly. (2000). United Nations Millennium Declaration. A/RES/55/2. UN.

KEY TERMS AND DEFINITIONS

Award Criteria: Award criteria enable the contracting entities to evaluate and compare the different bids received based on price solely or on price and quality, the criteria shall be predetermined, set in advance, and linked to the subject matter of the contract.

Contract Performance Clauses: Contract performance clauses are included in the contract that has to be signed between the contracting entity and the winning bidder and specify the way in which a contract should be carried out, these clauses can include green considerations.

Green Products and Services: The adjective "green" referred to products and services indicates that such products and services require the use of fewer resources, have a fewer impact on the environment, are designed to last longer when compared to similar ones.

Life-Cycle Costing (LCC): LCC corresponds to the sum of the costs that the contracting entity bears during the whole process of procuring, owning and disposing, LCC calculation includes purchasing price and the associated costs, operating costs and end-of-life costs.

Performance-Based Specifications: A performance-based specification usually describes the desired results and outputs of the procurement exercise and indicates how they will be measured; this type of specification does not include the inputs to be provided or the methods to be used by economic operators.

Sustainable Procurement: Procurement is called sustainable when it combines requirements, specifications and criteria that encompass the environmental, social and economic dimensions of sustainability.

Technical Specifications: Technical specifications have the function to provide a detailed description to the market of the good, work, or service to be procured, thus allowing interested economic operators to draft appropriate bids; they shall be proportionate to the size and complexity of the procurement.

Whole-Life Costing (WLC): Life-cycle costing that takes also indirect costs into account is referred to as whole-life costing (WLC).

Chapter 3
Green Audit for the Environmental Sustainability

Ganga G.
Sree Ayyappa College, India

ABSTRACT

Green Audit works as an environmental consultancy and reviews organizations with the aim of monitoring the performance of companies and organizations whose activities might threaten the environment and the health of citizens. Green Audit is to give citizens the information they need to be able to question when an institute or organization is destroying the environment which we all depend on. Most of the public and private organizations do not have suitable strategies to inspect or suggest the control measures be taken to avoid or reduce such environmental risk, which provided the impetus for the establishment of Green Audit. Green Audit undertakes and supports independent studies in the general areas of environmental public health, pollution, energy efficiency, and social and policy research in relevant areas. It has become increasingly clear that the scientific measurement and appraisal of human health risks, including pollution, are in the hands of the government and establishment-controlled organizations.

DOI: 10.4018/978-1-5225-7083-7.ch003

INTRODUCTION

Green Public Procurement (GPP) is a process in which public sector pursue to reduce the negative environmental impacts caused by the purchasing of goods, services, and other public related works. The drastic effects of environmental degradation are steadily increasing — natural resources getting limited which is going to adversely affect the further growth of a nation and the future of the next generation. Increasingly, the world's environment is under threat from unsustainable forms of development. Implementing Green policies and making it as practice can limit the further depletion of the environment and thus the available resources can be protected. Insisting procurers that are legally mandated to buy green, practicing and executing is to be made mandatory.

Getting started on green procurement in public sector can be challenging, due to the existing procedures tend to be formal, arduous, involves coordination among multiple divisions and reporting hierarchies, and budget flexibility are limited (Perera, 2011). Most of this burdensome administration procedures are and time is lagging ceremonious processes that have not been revisited for many years. Such a situation can be mitigated by the implementation of modernizing procurement laws and procedures that allow e-procurement. Green purchasing is also about influencing the market. This can be made possible by promoting the use of green products. To enhance the movement of such products public authorities can provide the industry with attractive incentives for developing green technologies and products. Government is to take new initiative or call project proposal from Government R&D labs, Institutions, Educational projects for developing new ideas and avenues. Financial aids and subsidies should be made able for the implementation of such proposals. Intensification of multidisciplinary research on this aspect needs to be carried out to develop green functioning, and Green products In India Organic farming and Green Product acceptance should widely be enforced among the public. Thought provoking campaigning programme through media and channels and social networking site can evoke waves in Public minds. Green Procurement in some public sectors, like Public transport and construction, Public health services and public education purchasers command a significant share of the market for a nation, and so their decisions have considerable impact. Decision made by domestic consumers play a pivotal role is marketing. Hence intense awareness at village, Gramapanchayath level, and other women empowerment sectors are needed. Interference of the Government in fixing the values of the value-added product is needed to protect farmers and their produce. Nature conservation and protection act and laws should be amended in such a way to promote the use of green products without exploiting nature. The green audit should make a mandatory programme in all public and private sectors. Violation of rule against Green policies should seriously be considered penalty and fine should be implemented for such cases. Sustainable development has emerged as a new paradigm of development (UNDESA. 2002), which is reciprocal to the current scenario of development that over-exploits natural environment.

The green environment is a fundamental part of sustainable development and growth for every nation's prospects. There is a correlation between the world's environment - the physical, biological and social surroundings and their interactions that sustain life and the objectives of human development — a sustainable, eco-friendly approach in all sectors in need of an hour for the future of a better nation. Hence, to attain environmental sustainability there is a need to increase the institutional, human resource and technological capacities of developing country governments, organizations and communities to plan and implement development policies, programs and activities that are environmentally sustainable.

Insisting procurers that are legally mandated to buy green, practicing and executing so becomes so much easier. It legitimizes the time, effort and money spent by procurers on consulting with environmental specialists, suppliers and stakeholders, and designing specification and award criteria that will crowd in the critical multiplier benefits: the broader commercialization of green goods/services across the domestic market (Perera, 2011).

WHY THIS GPP IS NEEDED?

The green environment is a fundamental part of sustainable development and growth for every nation's prospects. There is a correlation between the world's environment - the physical, biological and social surroundings and their interactions that sustain life and the objectives of human development. The drastic effects of environmental degradation are steadily climbing. Depletion of natural resources is limiting the development potential of developing and industrialized nations alike. Increasingly, the world's environment is under threat from unsustainable forms of development.

All nations must work together to solve environmental threats of a global nature and those who undermine sustainability at more local levels. Industrialized countries need to reduce their levels of consumption and continue to improve the management of their natural resources. Developing countries continue to seek development cooperation with industrialized countries, to address their sustainable development priorities. Each country must play its legitimate role, based on the principle of mutual benefit and obligation, and according to its relative technical and financial capacities. This can be achieved various methods, of this implementing eco-friendly public procurement system is one of the significant priority areas which may bring social responsibility among citizens.

Though there are limiting factors such as poverty, the pressures of everyday needs, inequitable control over resources, coupled with differing attitudes and levels of knowledge about the environment on the part of both individuals and societies, at the other end of the spectrum, there are macroeconomic pressures such as low commodity prices and debt burdens.

Other factors are working against the environmental interest: inadequate financial and human resources, underdeveloped institutional and technological capabilities, small opportunities for people to participate meaningfully in the development process, inappropriate or narrow economic and social policies, and inadequate incentives for environmentally sound behavior (Glosbe, na).

What Are the Challenges for the Implementation of GPP?

The challenges for the implementations are (Qiao & Wang, 2011):

- Green products perceived to be more expensive
- Lack of environmental knowledge
- Lack of managerial support
- Lack of tools and information
- Absence of a single uniform approach for defining criteria
- Lack of training Lack of mechanism monitoring the systems
- Absence of incentives or subsidies for the implementation of products
- Lack of sufficient suppliers

Lack of Research in Establishing Technologies Related to Green Procurements

For the promotion GPP of there are few best practices to be adopted (Buying Green, 2016; Alejandre, et al., 2010)

- Green public Procurement Legal and Policy framework
- Understanding Market Capacity and Assessing cost-benefit
- Introducing Environmental Standards in Procurement
- Professionalizing Green Public Procurement
- Raising Awareness
- Monitoring of the Green Public procurements
- Identify high demand area in Public sectors and strict implementation policies to be made in those sectors like public transport and construction, health services and education

- Intensifying of Research and development in the Public sector such as R&D Labs, Research institutions, Universities, etc.
- Green Audit is to be made mandatory for in Public and Privets sector which include Govt Institutions, Public services, and Education system and Banking sector, etc,
- Educate the domestic users
- Promote legal volunteers
- Create awareness at Primary Education system onwards.
- Change in Lifestyle to promote green habits.

GPP FOR SUSTAINABLE DEVELOPMENT

Socially and environmentally sustainable public procurement is always beneficial for the society and the participants in the markets. This may provide support to contracting authorities, entities and suppliers. Sustainable development has emerged as a new paradigm of development which is reciprocal to the current scenario of development that over-exploits natural environment (Shannuganthan & Warren, 1997).

To achieve environmental sustainability, there is a need to increase the institutional, human resource and technological capacities of a nation. Governments, organizations, and communities should plan and implement development policies, programs and activities that are environmentally sustainable. Strengthen the capability of a nation like India to contribute to resolving global and regional environmental problems. This can be accomplished along with meeting the nation's developmental objectives by efficiently integrating environmental considerations into decision-making. The framework GPP measures need a policy well suited for the local and regional realities of each organization. For sustainable development an adequate training of the public bodies and the monitoring of the GPP performance is essential.

For example in the case of a Banking sector, sustainable green policy implementation should be environmental regarding in friendly way emissions and pollutions. However, when compared to other industries internal environmental impact of the banking sector such as the use of energy, paper and water are relatively low and clean. Environmental impact in banks (Ilag, na) is not physically related to their banking activities but with the customer's activities. Customers and employees must be educated, and care should be given on to avoid humanmade environmental impacts. Moreover, environmental management in the banking business is like risk management. It increases the enterprise value and lowers loss ratio as higher quality loan portfolio results in higher earnings. Thus, encouraging environmentally responsible investments and prudent lending should be one of the responsibilities of the banking sector. According to Indira (2016) Green Banking and green policies, implementation will be mutually beneficial to the banks, industries and the economy. Not only "Green Banking" will ensure the greening of the industries, but it will also facilitate in improving the asset quality of the banks in the future."

For the eco-friendly sustainable development of a nation purchasing the green product and use of green products should make mandatory. This will, in turn, beneficial for influencing the market. To enhance the movement of such products public authorities can provide the industry with attractive incentives for developing green technologies and products. The government should take new initiative or call project proposal from Government R&D, institutions Students projects for developing new ideas and avenues. Financial aids and subsidies should be made able for the implementation of such proposals. Intensification of multidisciplinary research on this aspect needs to be carried out to develop green functioning, and Green products In India Organic farming and Green Product acceptance should widely be enforced among the public. Thought provoking campaigning program through media and channels and social networking site can evoke waves in public minds. Green Procurement in some public sectors, like Public transport and construction, Public health services and public education purchasers command a significant share of the market for a nation, and so their decisions have considerable impact. Decision made by domestic consumers play a pivotal role is marketing. Hence intense awareness at village, Gramapanchayath level, and other women empowerment sectors is needed. Interference of the Government in fixing the values of the value-added product is needed to protect farmers and their produce. Nature conservation and Protection Act and laws should be amended in such a way to promote the use of Green products without exploiting nature (NAPP, 2015 & Green Procurement Standards, 2012).

SCOPE OF GPP AND ITS IMPLEMENTATION

Green public procurement is an essential instrument in achieving socio-political objectives. Incorporating environmental considerations and social considerations into procurement practices contributes to sustainable development and improve national environmental-quality objectives (Buying Green, 2016) Services of GPP at various sections are given in Table 1.

Table 1. Services of GPP in various sectors

Area	Services
Implementation Strategy	Provides guidance on programming options for environmental sustainability and to carry out program initiatives
Human Resources	Use of existing environmental expertise and actively support the development of additional environmental and sectoral expertise both within and outside the Public sector. Additional skill acquisition programs and training programs working people will help to promote the enhancement of human resources for the implementation of the policy effectively.
Economic resources	Promote new environmental initiatives through reallocation of funds within those programs where the environment is determined to be a priority area. Partnership with an International organization will promote economic support.
Partnership	Develop a partnership with organization/national which support environmental protection.
Support Private/lateral Institutions	Advocate and support measures to be provided to multilateral institutions to incorporate environmentally beneficial policies and practices and to modify their programs where necessary to address environmental issues.
Performance in the Field	Performance-based appraisal for employees and organization for green practicing.
Training and Sensitization	Design and implement a comprehensive training program on environmentally sustainable development for its managers and staff, including its field officers.
Information and Analysis	Adopt data management systems to allow data on environmental programs and projects to be readily accessed and will analyze and disseminate information on environmental issues.
Public Outreach programs	Foster awareness and involvement in environmental issues in public.
Internal Operations	Ensure that its internal operations such as environmental stewardship, measures for energy conservation, material recycling waste minimization and procurement environmentally safe materials are consistent.

ADVANTAGES OF GREEN PUBLIC PROCUREMENT

Sustainable Green public procurement is an essential mechanism for a nation to achieve environmental objectives (Bouwer et al., 2005; Green Public Procurement, 2015; the European Ecolabel, 2008)

Environmental Benefits

Sustainable public procurement can contribute to solving environmental issues like:

- Emissions of greenhouse gases (low environmental footprint during their life-cycle).
- Water usage (through water saving).
- Energy efficiency and resource usage (through procuring eco-friendly products).
- Air-, water- and soil pollution (limiting the use of toxic substances).
- Waste (through procuring re-used materials or recycled).
- Sustainable agriculture (through procuring organically produced food).
- Deforestation (through procuring wood and wooden products from legally felled forestry).

Social Benefits

Getting public acceptance for such public procurement system is the significant milestone for the implementing GPP. Better understanding among the public is needed about its role in contributing social benefits in national as well as an international level. This can achieve by addressing relevant social requirements that might lead to an improved labor market through requirements on employment and social integration.

Influencing Private Consumers

Sustainable public procurement can establish the norms for the private sector as well as for society at large, thereby affecting the market. Through developing a sustainable public procurement strategy, communicate its initiative and result, procuring organizations can be able to show its possibilities that can lead to positive results (NAPP, 2015).

Create Awareness About Environmental Issues

environmental issues (NPPA, 2015). Understanding the adverse effect of non-Ecofriendly products/by-products will promote the production of Eco-friendly products. This will also highlight the benefits resulting from environmentally friendly alternatives.

To Increase Equality in Society at Large

Similar good and products with the same quality can be provided to society at large. Through social requirements, one can make sure that the public sector acquires goods and services produced ethically through using requirements on non-discrimination between sexes or differences in ethnic and cultural background.

Economic Benefits

Procuring an efficient cost-saving, energy-saving product needs for the society with a costing mindset. One example for this is procuring energy-efficient IT where money can be saved both through lower energy usage and by simplifying reuse/recycling at the end of life. Goods and services with better efficiency often have a higher purchase price but will generally allow for the organization to save money in the long run.

Innovate and Implement New Strategies

The demand for sustainable green practices might give signals to the market where there is a need for new strategies, which in turn might provide an incentive for developing new goods and services.Replacing and implementing such green procurement policies with old approaches need innovation and ideas. Applying sustainability to the public procurement process might lead to an essential incentive for the industry to develop "green" technology and goods *(NPPA, 2015).*

Contribute to a Circular Economy

Today the norm is a linear economy where used products end up at the waste ground. This contributes to an impoverishment of resources (NAPP, 2015). In a circular economy, the focus lies instead on minimizing resources going to waste by simplifying for recycling, deconstruction and re-using. Sustainable public procurement can contribute by procuring more services instead of purchasing products and through this contribute to the development of new business models.

Lower Consumer Prices for Eco-Products

When introducing sustainable criteria in procurement, a possible outcome is to affect the market and contribute to new actors within the field of "green" technology and products. This, in turn, might lead to increased competition and lower prices in the long run (NAPP, 2015).

Health Related Benefits

To Improve Quality of Life

Implementation of GPP will promote improved service to society and through this increased quality of life. For example, cleaner public transport will improve air quality. Less toxic chemicals in cleaning products will result in a healthier working environment (NAPP, 2015).

APPROACHES TO IMPLEMENT GPP

Sustainable public procurement public authorities meet their needs for goods and services in a manner that secures a good affair with respect to the entire life-cycle. This while considering benefits not only to the organization but also for society at large and minimizing damage to the environment (NAPP, 2015). To make substantial progress towards environmental sustainability, three concepts will need to be better understood and applied. The three concepts are (Boromisa, Tišma, & Anastasya, 2016):

1. Understanding the Ecological Basis for Development
2. Assessing the Economic Value of the Environment
3. The complicated Interrelationships of Poverty, Population and Natural Resources

MAJOR GREEN PROCUREMENT STRATEGIES FOR AN ORGANIZATION TO GO GREEN

The few steps which can be followed for a sustainable green procurement for any public/private organization (Buying Green, 2016; Green Public Procurement in Europe, 2005; Green Procurement Standards, 2012; Marron, 2014 and Pearce, et al.,1989):

1. Commit to Being Green. Make everyone's responsibility and make it is mandate and create a green policy.
2. Identify and Categorize Your Needs. The first step is to develop green specifications and standards for every product.
3. Establish Green Selection Criteria and Their Impact on Award Decisions
4. Once you have identified the standards you are going to use for each category, you have to outline the selection criteria, weight and prioritize them, and figure out how much of an impact they are going to have on your decision
5. Focus on Identifying Products and Services which are Green
6. Choose always a green supplier. Make it a policy that, unless you are not able to identify at least three green suppliers, that a non-green supplier should not even be allowed to enter a bid.
7. Always Use a Life-Cycle Costing Approach
8. Consider the total environmental impact of the product you are considering from the harvesting and processing of the raw materials through its final disposal before making a decision. Be sure to include efficiency, waste, recyclability, and material composition into your analysis
9. Include Green Performance Clauses in Every Contract
10. Incorporate clauses into every contract that allow you to enforce penalties or terminate the contract if the supplier does not meet the minimum green and sustainability requirements that they commit to.
11. Communicate and Inform
12. Policies and practices, needs to be communicated with buyers so that they can differentiate the products that are truly green from those that are coated in green washing
13. Use of Green Technology.
14. Use E-procurement, E-sourcing, and other E-Systems, run on energy-efficient technology, to buy online rather than using reams upon reams of paper that result in the unnecessary destruction of forests to research, contract, and buy products and services
15. Make it Easy
16. Design every policy, process, and system developed and deployed in support of green to be easier to use than the alternatives

FUTURE PROSPECTS

Implementation of this policy is an on-going process. Internal and external capabilities will need to be built up, and the complex concepts underlying environmentally sustainable development will need to be further elaborated and tested (Doglas College, 2009). A comprehensive effort to integrate environmental considerations into the thinking and practice should be considered as a long-term integrated effort with Government and Public. Adequate amendments in policies are to be made on a need basis. Annual consultation and environmental organization, experts and the public are needed, and critical evaluation and reviewing need to be done before implementing the new policies.

REFERENCES

Alejandre, E., Traspaderne, V. A., & Ortiz de Elgea, V. A. (2010). Best practice on green or sustainable public procurement and new guidelines. Open House.

Boromisa, A., Tišma, S. & Anastasya. (2016). *Green Jobs for Sustainable Development*. London, UK: Routledge.

Bouwer, M., de Jong, K., Jonk, M., Berman, T., Bersani, R., Lusser, H., … Szuppinger, P. (2005). *Green Public Procurement in Europe 2005 - Status overview. Virage Milieu & Management bv, Korte Spaarne 31, 2011 A.J Haarlem, the Netherlands*. Available at http://europa.eu.int/comm/environment/gpp/media.htm#state

Collage, D. (2009). *Carbon Neutral Action Report*. Available at https://www2.gov.bc.ca/assets/gov/environment/climate-change/cnar/2008/ps/douglas_college.pdf

Glosbe. (n.d.). *Environmentally sound behavior*. Available at https://glosbe.com/en/fr/environmentally%20sound%20behaviour

Green, B. (2016). *A hand book on public procurement* (3rd ed.). Luxembourg: Publications Office of the European Union. doi:10.2779/246106

Green Audit. (2017). *What is Green Audit?* Available at: https://www.greenaudit.org/about/

Green Public Procurement. (2012). *A collection of good practices*. Luxembourg: Publications Office of the European Union. Retrieved from http://europa.eu/environment/gpp/)

Green Public Procurement and the European Eco label, Fact sheet. (2008). Toolkit developed for the European Commission by ICLEI - Local Governments for Sustainability, European Commission, DG Environment-G2, B-1049.

Indira C. M. (2016). Green banking practise in India. *International Journal of Advanced Research*, *4*(11). doi:10.21474/IJAR01/2156

Marron, D. (2014). Greener Public Purchasing as an Environmental Policy Instrument. *OECD Journal on Budgeting*, 3(4), 71-102. doi:10.1787/budget-v3-art23-en

NAPP. (2015). *Benefit from public procurement*. Available at: http://www.upphandlingsmyndigheten.se/en/sustainable-public-procurement/why-sustainable-procurement/beneficial-for-society/

Pearce, D., Markandya, A., & Barbier, E. B. (1989). *Blueprint for a Green Economy*. London: Earthscan.

Perera, O. (2011). *Procuring Green in the Public Sector: A checklist for getting started*. Available at: https://www.iisd.org/pdf/2011/procuring_green_public_sector.pdf

Qiao, Y., & Wang, C. (2011). Issues and Challenges in Implementing China's Green Public Procurement Program. *Journal of Environmental Protection*, *2*(08), 1034–1045. doi:10.4236/jep.2011.28119

Shannuganthan, P., & Warren, L. M. (1997). Status of Sustainable Development as a Norm of International Law. *Journal of Environmental Law*, *9*, 221.

UNDESA. (2002). *Global Challenge Global Opportunity: Trends in sustainable Development*. Available at https://sustainabledevelopment.un.org/content/documents/critical_trends_report_2002.pdf

Chapter 4
Green Public Procurement as a Leverage for Sustainable Development:
Documental Analysis of 80 Practices in European Union

Irene Litardi
University of Rome "Tor Vergata," Italy

Daniele Alimonti
Barcelona Graduate School of Economics, Spain

Gloria Fiorani
University of Rome "Tor Vergata," Italy

Andrea Appolloni
University of Rome "Tor Vergata," Italy

ABSTRACT

The chapter aims to highlight the advantages of GPP practices to address the environmental and economic problems, laying on the experiences of the European countries. The research has the objective to reconstruct the state of the art of green public procurement through the lens of a cross-country comparative analysis at European level. After introducing a systematic review of the literature and the core regulations of the GPP practice, the chapter underlines the results of a multidimensional analysis on a cluster of 80 GPP practices (sample updated to February 2014), identified in European Union and implemented by governments in 25 countries at central, regional, and local government level. The framework of analysis builds on several dimensions, mapping the main results on the following levels: geographic origin, the government level, GPP implementation period, main criteria for implementation, as well as environmental and economic impact of such practices.

DOI: 10.4018/978-1-5225-7083-7.ch004

INTRODUCTION

The absence of sustainable orientation of the current consumption model, based on increasing demand for resources, requires a change of direction, so that in the long term, the economic growth is not subject of failure and does not compromise the integrity of the environment and public health. In this scenario the role of public spending becomes important as it is able to direct the purchases and the market towards greater sustainability. It has been estimated that in OECD countries the amount of public procurement as a percentage of GDP in 2011, for the majority of OECD countries, is between 10% to 15%. The procurement instrument, with an average percentage of expenditure in *procurement* activities of 29% of the total public expenditure, confirms its pivotal role in the policies of the Member States, due to its ability to act as a driving force for the economy, and supporting the governments to achieve economic, social and environmental objectives (OECD, 2013). In the European Union, the public sector plays a key role in the economy, ranking first among consumers: in 2009, the annual expenditure on acquisitions reached two trillion euros (SEC (2011)853), approximately 19% of the GDP. According to most recent estimate it is around 16% of the GDP, with a decrease of three percentage points compared to earlier forecasts (European Commission, 2014), whilst maintaining the task of stimulating the research for more efficient technologies and facilitate developing of products that have less environmental impact. The ongoing research by the public authorities of useful solutions to curb their impacts on the ecosystem, along with increasing public opinion attention to the environment, have forced governments to pay attention, increase efforts and deliver resources towards sustainable development. This led to considering Green Public Procurement (GPP) as an instrument able of binding the environmental dimension to public spending, as well of changing the former consumption patterns among countries.

OBJECTIVES AND METHODOLOGY

The research aims to analyze the state-of-art of the main practices of Green Public Procurement (GPP) in the European Union. After introducing the framework of green public procurement theme as part of the main theories of reference in the process of modernization of the public administration (between the New Public Management (NPM) and Public Governance (PG) paradigms), a synthetic overview of the legislative framework is presented (Paragraph of literature and regulation of GPP).

The documental analysis (Bowen, 2009) on the last two decades regulatory measures and policy, initiated by the European Commission, aims to provide a summary of the major themes in the field of green public procurement, underlying the growing attention to the instrument by the Commission and its evolution towards Sustainable Public Procurement (SPP), along with the inclusion of social and environmental requirements, within the tendering procedures (focus of the Directives from the beginning 2014).

The empirical research part (Part of empirical studies) is based on an exploration of a multi-dimensional cluster of 80 GPP practices (updated in February 2014), identified by the European Commission starting with 2010 and implemented by the 25 Member States public administrations at local, regional and national levels. In particular, a systematic analysis was carried out on the published cards by the European Commission.

Each card is structured on six dimensions, and underlines:

- **The Objective of Procurement:** i.e. the reasons behind the public administration objectives to do a tender and the aims to be pursued;
- **The Background:** A summary of the already activated GPP activities by the concerned public administrations;
- **The Criterion Used:** The information on the environmental criteria adopted (falling into one or more categories available at Community level), as well as additional information about the tender;
- **The Results:** i.e. the description of the procedure outcome;
- **The Environmental Impact:** Results of the actions undertaken;
- **The Lessons Learned:** Gathering key points on which to set future strategies and identified mistakes to avoid in the future purchasing procedures.

The information contained in the sections of the cards have been systematized, summarized and elaborated in order to present a mapping of public administrations initiatives and their critical reading on the basis of the main theories and models (New Public Management, Public Governance, Triple Bottom Line/3BL). According to the theory of 3BL (Part of GPP literature and regulation review), the variables analyzed include, in addition to the geographical origin of the cases, the implementation period and level of government concerned, as well as the criterion used and the economic and environmental factors that are closely linked with the dimensions defining this theory.

In particular, each variable has been operationalized into indicators, such as: the starting year of the procurement contract; the country of reference; the complexity of the administrative structure, divided by local, regional/provincial and national; the tender sector (each containing the criteria to be used); the award criterion (most economically advantageous offer; lowest price or other methods); the presence or lack of economic and environmental impacts - positive, negative or none - both in quantitative and qualitative terms. Through the case studies analyzed it was possible to identify the overlap between the regulatory processes of the European Commission and the Public Administrations initiatives, the most active geographical areas, the main government level involved and finally, the most used environmental criteria and award criterion.

GREEN PUBLIC PROCUREMENT (GPP)

The GPP development in Europe since the early 90's, was driven by the key role played by Rio de Janeiro meeting (1992), which marks the birth of a compendium of directions to be followed in the twenty-first century by each State, namely Agenda 21. The latter implies, in particular, the need to review the purchasing policies of agencies and departments in order to improve, if possible, the environmental implications of public procurement (UNCED, 1992). The structure of the Agenda is divided into four parts: social and economic dimensions, conservation and resource management for development, strengthening the role of the most significant groups and implementation instruments.

The concept of sustainable development, defined in 1987, is also based on four key dimensions such as economic, social, environmental and institutional sustainability, and, such as Agenda 21, has some convergent points with the Triple Bottom Line (3BL) approach, introduced by Elkington in 1995. The author revises the concept of Corporate Social Performance (CSP) introduced by Carroll in 1979, which defines the achievement of the social goals of the businesses as the effect resulting from the combination of four different dimensions: economic, legal, ethical and philanthropic. Specifically, Elkington focuses its attention on the first three dimensions, placing at the core choices of Corporate Social Responsibility (CSR), the economic, social and, ultimately, the environment areas, as the key to success in the long run. European Commission, through the Green Paper of 2001 recognized the increased importance of corporate social responsibility. CSR policy is defined by business ability to integrate with activities, the environmental and social awareness, albeit voluntarily. Public sector sought a multi-stakeholder approach to the process of innovation and modernization of the public administration, aimed at the whole community, understood as a society, since the 90s. During this period the attempts were firstly to overcome the traditional public administration through the transfer of operating modes from the private sector to the public sector, in order to improve the government management (New Public Management - NPM, PUMA/ OECD, 1996). The difficult adaptation to these logics has led over time to a change in paradigm from NPM to Public Governance, which through a stronger cooperation between the public, private and civil society, focuses on the need for environment protection, the development of new technologies, the culture of individuality and internationalization of issues such as migration and development cooperation. Among the useful policy instruments to address the first requirement is the GPP, which has had a remarkable development, both at legal and theoretical level (Ricotta, 2012; Becchetti, Bruni & Zamagni, 2010; Bertoldi, Damian Bornás, Monni & Piers de Raveschoot, 2010; Bettini, Masini & Ricotta, 2006; Saltari & Travaglini, 2005; Szuppinger, 2005), to become a pillar on which it can support the environmental strategies at European level (Tukker et al., 2008). Reviewing the main regulations stages on the theme of green procurement, 1996 definitely marks a turning point, along with European Commission's Green Paper on public procurement, which introduced the environmental and social requirements at the level of national and communitarian competitions.

In 2001, the Sixth Environment Action Program of the European Community (COM (2001) 31), defines the Integrated Product Policy (IPP) as a new paradigm of growth which is able to create wealth and competitiveness on the basis of eco-environmentally friendly products. Furthermore, is suggested a strategy for fostering the production policies, requiring less use of resources, therefore, inducing a reduced environmental impact, ecological risks as well as greater attention to the production of waste, starting from the design phase. All these aspects contribute to the life cycle approach of the product, which along with the stakeholder's engagement, plays a key role in fulfilling the objectives. The bottom-up engagement model of this policy converges fully with the last decade slogan initiated by the Agenda 21 - "think globally, act locally". Later, with the National Action Plan on GPP, the IPP instrument is inextricably linked to the theme of green public procurement, stipulating within the tenders the integration of environmental criteria, thus ensuring increased widespread among the Member States (COM (2003) 302). The European Commission (2014) reveals that twenty-two countries have included in their strategies and published their national plan of action, also sometimes in a renewed version, while, the other six do not possess any documents amenable to this plan (Table 1).

Table 1. Update of NAP GPP EU28 implementation by EU member states

	Countries
NAP or equivalent	Austria, Belgium, Bulgaria, Cyprus, Denmark, Finland, France, Germany, Ireland, Italy, Latvia, Lithuania, Malta, Netherlands, Poland, Portugal, United Kingdom, Czech Republic, Slovakia,
Without NAP	Croatia, Estonia, Greece, Luxembourg, Romania, Hungary

Source: European Commission, 2014

In the same period, due to the considerable complexity reached by the EU administrative law there are challenges to providing new solutions into the public procurement strategies, referring to regulatory provisions in the support of the environment. In a first stage, the Communication 274 of 2001, with which the Commission renew previous publications, describes the entire procurement process, identifying the contracts included in the directives and highlighting the possibilities offered by existing rules to structure the various stages of a procurement tender, starting with the definition of its subject to the execution and completion of the same. Later, the Directives 17 and 18 of 2004 support further certain decisions of the European Court of Justice, filling some gaps in legislation and providing the legal support on GPP. Both Directives, guiding the Member States currently, respectively govern the procedures of "Public procurement in water, energy, transport and postal services" and "Public procurement in work, supply and services". In support to the Member States efforts, the European Commission enacted the Communication 400 (2008) "Public procurement for a better environment", which accompanies the action plan "Sustainable Consumption and Production" and "Sustainable Industrial Policy" of the same year (COM (2008) 397).

According to the first directive, the Green Public Procurement is defined as "the approach by which Public Administrations integrate environmental criteria into all stages of the acquisition process, encouraging the spread of environmental technologies and development of environmentally sound products, through research and choosing outcomes and solutions that have the least impact on the environment throughout the life cycle". The definition, widely accepted by the literature, recognized the GPP as a tool able to guiding the market for products and services and allowing the Public Administrations to carry out their purchases with less impact on the environment, through the adoption of minimum environmental criteria. The second directive is concerned with the implementation of a several measures aimed to improve the energy and environmental performance of the products throughout their life cycle and to stimulate the demand and the consumption of better-quality products, creating a "virtuous circle". In the midst of financial and economic crisis, the European Union redefines its long-term objectives with a prospective to achieving sustainability. The objective is to "turn the crisis into an opportunity to promote the financial and ecological sustainability, while developing likewise, a dynamic society with low-carbon economy, that makes efficient use of resources, both knowledge-based and socially inclusive, and to support this approach at the global level" (COM (2009) 400).

In March 2010, European Commission launched the Europe 2020 Strategy, which, in line with the provisions of the Kyoto Protocol, providing three approaches of growth (inclusive, smart and sustainable) to be implemented in the Member States, reaffirms the importance of public procurement as an element to "foster the transition to efficient economy in terms of resources and low-carbon emissions" (COM (2010) 2020). The Strategy is inserted, in turn, into a larger project of long-run in which the European Union is involved, the "European Roadmap", which requires the transition to a low carbon economy by 2050: -25% in 2020, - 40% in 2030, -60% in 2040 and -80% in 2050. In reaching targeted objectives, there is a need of a paradigm shift in the public authorities' culture, that, enforcing the support to the future strategies and European Union norms (COM (2011) 896), will be able to substitute, at the criteria of policy assignment of the "lowest price" to that of the "lowest cost", considering the life cycle of the procurement (European Commission, 2014). The life cycle of the environment, refer to the concept of Life Cycle Assessment (LCA), linked also to the Life Cycle Costing (LCC), which is the core of the Green Public Procurement (European Commission, 2003; COM (2003) 302). A Life Cycle Assessment is defined by the European Commission, as a method of assessment of the product, process or activity impact on the environment throughout life cycle. The public procurement norms identify a set of contracts, including those containing environmental considerations, seen as "particularly complex", providing two alternative methods of assessment: lowest price and most cost-effective offer. In the second stage, defined by the situation in which the client undertakes a competitive dialogue phase, can be used for the costs calculation the criteria of Life Cycle Costing, a tool that allows to consider the entire life cycle of the works, services or supplies covered by the procurement and not just their price. The traditionally used method by businesses and governments to calculate the life cycle cost is based on a purely financial analysis with four dimensions: cost of investment, management, services and final disposal. The LCC method proposed within the environmental strategies, adds a fifth dimension, such as the external environmental costs, calculated by assigning a monetary value to the externalities detected by the analysis of the life cycle of the product (European Commission, 2014).

In conclusion, the latest EC policies requires an increasingly strong orientation towards transparency, competition, reduction of consumption and likewise, an economic and socio-environmental return as stated by European Commission Directives "on the adjudication of the concession contracts" (2014/23/EU), in the matter of "procurement and which revoke the Directive 2004/18/EC" (2014/24/EU), "on procurement procedures of the service delivery entities operating in water, energy, transport and postal services and which repeal the Directive 2004/17/EC" (2014/25/EU). Through these directives on public procurement, the EU Parliament tries to simplify the structure of the purchase contracts, in order to include requirements fostering social and environmental responsibility. One of the directive objectives is to guide the public authorities toward specific social goals at the national and international level.

The recent developments raise, therefore, increasing attention to the social dimension within procurement and its subsequent transition towards Sustainable Public Procurement (SPP), a broader concept of GPP, meaning the government's attempt to strike the right balance between the three pillars of sustainable development - economic, social and environmental, in all phases of the procurement purchase of goods, services and works, (3BL approach), with the objective to integrate competitiveness and sustainability within the supply chain, along with the attention thorough the process, to human and workers' rights. Given the maturity of the Green Public Procurement practice, the European Commission (2004) underline its benefits and challenges (Table 2).

EMPIRICAL RESULTS ANALYSIS

A diversified framework of practices is observed through the multidimensional analysis of the 80 GPP practices identified by the European Union. Specifically, concerning the research first dimension, namely the geographic origin, it is possible to observe that, although 33 cases of the total sample are from the Green-7 (PricewaterhouseCoopers, Significant and Ecofys, 2009), diverse GPP front-runners countries in Europe are under-represented in the below rank (Table 3).

Table 2. GPP: Benefits and challenges

Benefits	Challenges
Environmental	• Lack of political support
• Allows public authorities to achieve environmental objectives	• Perception of higher costs associated with green products
• Example for private consumers	
• Increases awareness to environmental issues	
Social and Health	• Lack of legal expertise in the application of environmental criteria
• Improve the quality of life	
• Support for high standards of environmental performance for products and services	• Lack of practical tools and information
Economic	• Need for systematic implementation and integration into management systems
• Allows to save money and resources when considering the life cycle cost	
• Provides incentives for industry to innovate	• Lack of training
• Reduce prices for environmental technologies	• Lack of cooperation between authorities
Politicians	
• Effective way to promote the public sector commitment to environment protection and sustainable consumption and production	• Limited established environmental criteria
	• for products and services

Source: European Commission, 2014

Table 3. Geographic origin of the cases analyzed

Country	Number of cases
Spain, United Kingdom	8
France	7
Germany, Sweden	6
Italy	5
Netherlands	4
Austria, Belgium, Denmark, Finland, Ireland, Malta, Slovenia	3
Bulgaria, Norway, Portugal, Hungary	2
Estonia, Greece, Iceland, Latvia, Lithuania, Romania,	1
Total	**80**

Source: Author's elaboration

It should be noted that only 29 cases out of 80 are from Mediterranean countries such as Spain, France, Italy, Malta, Slovenia, Portugal and Greece; in particular, the first three accounts for about 70% of this subset.

The second dimension analyzed was the levels of government mostly engaged to activities of GPP (Figure 1).

Such indicator underlines the predominant engagement of the local government (55%) (Municipalities, universities, schools, museums and other organizations or local authorities), which, often supported by political level, are engaged in environmental causes since early 2000s (Barcelona (Spain), Budapest (Hungary) and Dunkerque (France) or, more rarely, from the 90s, as in the case of the city of Esbjerg, Kolding (Denmark) and Stockholm (Sweden). The 25% of best practices comes from the regional/provincial level; several cases highlights the engagement of regional commissions of Spain, Belgium and Italian, such as the Department of the Environment and Territorial Policy of the Basque Government (IHOBE), the Regional Agency for Environmental Protection of Tuscany (ARPAT), the Lombardy

Figure 1. Levels of government in the cases examined

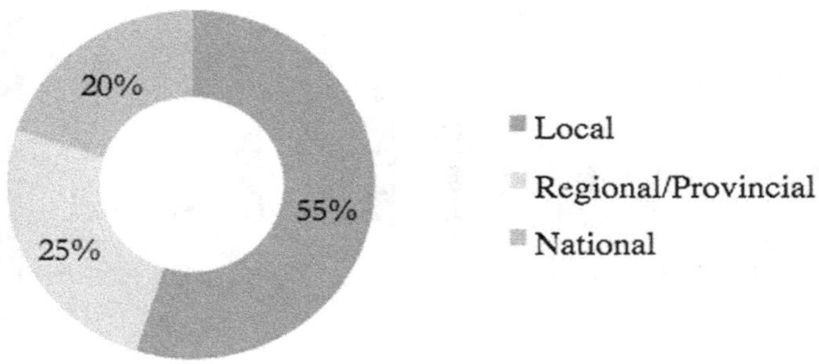

Central Regional Agency of Purchases and Valle Aosta Region (the last active on environmental issues since the mid-2000s) and some County Councils in the United Kingdom (Gloucestershire, Cornwall and East Ayrshire). The last 20% is represented by the national public administration, such as ministries (Hungary, Latvia and France) and commissions (Italy, Ireland, Lithuania, and Germany), introducing environmental criteria with the aim to encourage its application to lower government levels and facilitate their introduction within the procurement tenders.

In terms of reference year of implementation, all cases covers the period 2004-2013 (Figure 2), with a peak in 2009 and 2010 (40%), which coincides with the EU enactment period of most significant regulatory actions and policy in terms of GPP, including: COM(2008) 397 "Sustainable consumption and production" and "Sustainable industrial policy", COM(2008) 400 "Public procurement for a better environment", COM(2009) 400 "Mainstreaming the sustainable development into EU policies: 2009 Review of the EU Strategy for Sustainable Development", COM(2010) 2020 "A strategy for smart, sustainable and inclusive growth".

The criteria covered by GPP practices selected by the Public Administrations are essentially those promoted by the European Commission and links to the enlisted areas introduced by Table 4, and just in a few isolated cases are identified ad hoc national criteria.

Among the 80 cases of procurement the key priority sectors identified are: Construction, Copying and graphic paper and Transport as per, respectively, 18, 14 and 11 cases; followed by Electricity and Cleaning products and services (9 cases), Office equipment (8 cases), Furniture (6 cases), Food and Catering services (5 cases), Street lighting and traffic signals and Gardening products and services (4 cases), Wall panels and Textiles (3 cases), Water management and Cogeneration (2 cases), Hard-floor coverings and Roads construction and traffic signs (1 case); while, remaining sectors are not mentioned in any case practice examined.

Figure 2. Sampled cases distributed by year of reference

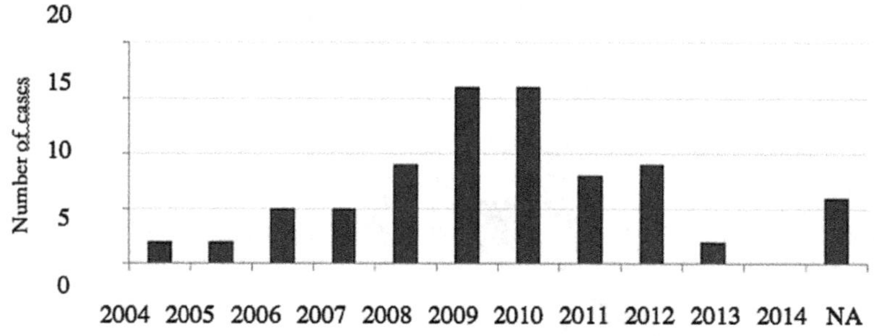

Table 4. The priority GPP sectors by the European Commission

• Equipment of imaging	• Wall panels
• Electric equipment used in the health sector	• Hard- floor coverings
• Office IT equipment	• Gardening products and services
• Furniture	• Cleaning products and services
• Copying and graphic paper	• Roads construction and traffic signs
• Cogeneration (CHP)	• Waste management
• Construction	• Water Efficiency Labelling
• Electricity	• Food and Catering services
• Street lighting and traffic signals	• Textiles
• Thermal insulation	• Toilets and urinals
	• Transport

Source: European Commission, 2014

The European Commission emphasizes the need to distinguish between two categories: "core" and "comprehensive" criteria. The first criteria are designed to allow easy application of GPP, focusing on the specific area of the product environmental performance and aimed at keeping the administrative and economic costs; while the second criteria take into account more aspects or higher levels of environmental performance, with consequent higher costs. In fact, the first should be based on the European Eco-label criteria, which address the main environmental impacts and are easier to fulfill, while the second resume additional criteria of the Eco- label that can be considered important for the definition of a specific product. Considering that 21 out of 80 cases analyzed are multi-sector procurement GPP (Figure 3), of which 9 include the "product scheme" which have not been reported by the EU Commission as a sector, in the research has been considered as such because of its recurring presence. On a similar vein is the particular case of the construction, office equipment and furniture sectors.

Figure 3. The number of EU GPP sectors within cases examined

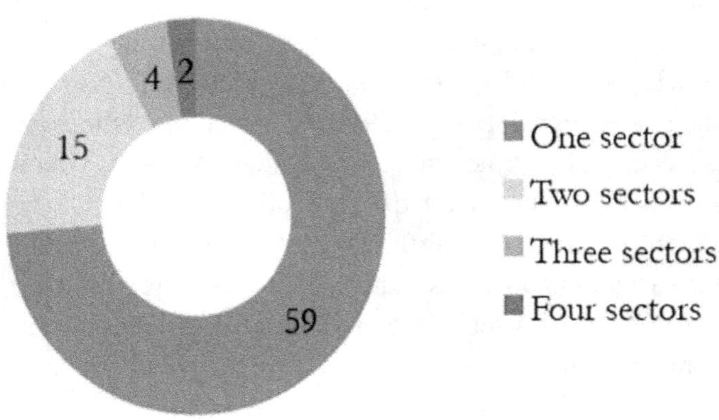

Such assumption raises the complexity of the above mentioned sectors and therefore the support of a "product scheme" specification, as the Commission itself points out in the Directive COM(2008) 400, along with the "ten priority sectors" for GPP, that have been selected on the basis of the importance of the relevant sector in terms of the scope for environmental improvement; public expenditure; potential impact on the supply side; example setting for private or corporate consumers; political sensitivity; existence of relevant and easy-to-use criteria; market availability and economic efficiency, such as: Construction (covering raw materials, such as wood, aluminum, steel, glass, as well as construction products, such as windows, wall and floor coverings, heating and cooling equipment, operational and end-of-life aspects of buildings, maintenance services, on-site performance of works contracts), Food and catering services, Transport and transport services, Energy (including electricity, heating and cooling coming from renewable energy sources), Office machinery and computers, Clothing, uniforms and other textiles, Paper and printing services, Furniture, Cleaning products and services, Equipment used in the health sector (European Commission, 2008). Not having found a high repeatability between the multi-sector associations is not possible to affirm a particular behavior in the last years, by the public authorities, during the tenders.

By and large, within the enlisted case samples analyzed, it is possible to identify many features that characterize Green Public Procurement. Firstly, there is the use of criteria and eco-brands, based on the Life Cost Assessment that assures the inclusion of environmental aspects in the tendering products throughout the entire life cycle, from the extraction to the end life of the product. Secondly, in respect of the competition, in the application of public procurement directives, the PA is given free options to promote environmental protection, through the introduction of specific technics in the procurement tender documents. It constitutes a barrier to participation in the tendering competition but also an innovative stimulus to competitors; examples are the minimum environmental criteria and the eco-brands, particularly present in the 80 cases. To this, is added the consideration of Life Cycle Costing, applied in some of the case projects examined, with the aim to achieve cost reduction and the assessment phase of the offers prices received.

Within the schemes, yield information is also referring to the criterion award used during the tender. In most cases (39) has been adopted the award mechanism based on most economically advantageous tender (MEAT), which suppose the simultaneous evaluation of the price and other quality and technical aspects, including the environmental dimension. Among the 25 cases examined, the documents reviewed do not report information about this indicator, in 9 cases refers to the lowest price, and finally in 7 cases, the PA make use of alternative methods that not necessarily relates to the previous ones.

Figure 4. Award criterion in the cases examined

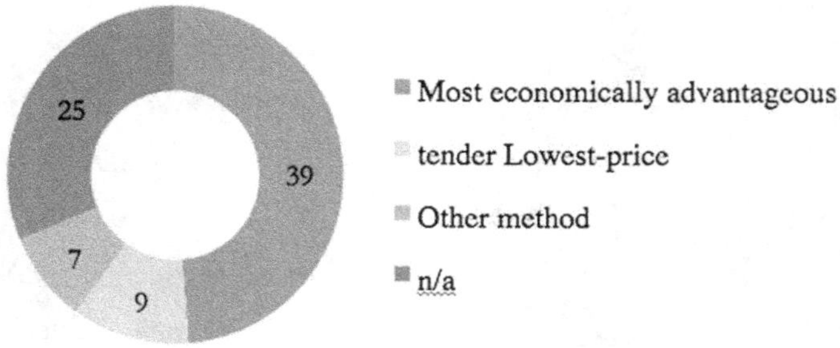

The criterion of most economically advantageous tender has allowed various public entities, through a fair approach to weighting the score, to reach a good compromise between price and quality. In fact, this method has awarded, in procurement, the competitive actors with respect to the environmental policies but weaker in terms of the economic proposal. The use of most economically advantageous tender (MEAT) helps to ensure the compliance of core requirements for the award procedure, such as: the connection between the criterion award and subject of the contract, a limited choice options for the PA customers, prior notification of the award criterion within the tender documents, the distinction between selection and award criterion and, finally, the respect of EU law (European Commission, 2011). In terms of the economic and environmental impacts was necessary to differentiate the results of the qualitative and quantitative cases, distinguishing the presence of information and numerical data. In terms of economic impact, in 41 cases out of 80, there is not information about financial results; of which 50% of cases are Mediterranean countries. In the left 39 cases, in 10 cases there was a costs reduction (quantified in euros) or in reduced percentage, comparatively to previous concluded contracts by traditional procedures (4 of which belongs to Mediterranean countries). In 16 cases, the results in terms of positive economic impacts are substantiated in a qualitative description of the achievements. In 6 cases has been observed no variance of costs compared to previous contracts while, in 7 cases, the use GPP has resulted in an increase of costs, quantified in 5 cases.

In terms of environmental impact, worth noting the positive contribution recorded in almost all cases studied (77 cases out of 80), such aspect being the central element of European Commission published cards. In particular, 33 cases, of which one-third belonging to Mediterranean countries, are distinguished by the inclusion of quantitative details in terms of reduction of CO_2 emissions (in tons), and greenhouse gas (GHG), electricity use (in kWh), water saving (in litres). In the other 44 cases, there is the generic description of the environmental benefits by linking the importance of actions promoted by European Commission about

Figure 5. The economic impact in the cases examined

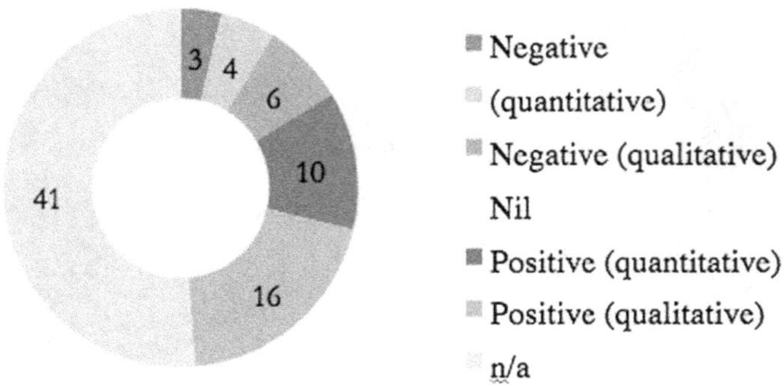

the relevance of the environmental impact. Among them, 40% cases belong to the Mediterranean countries.

CONCLUSION

From the analysis of regulatory measures and policies promoted by the European Commission is observed a high degree of maturity and attention to green public procurement. In line with the context of the analyzed cases, research findings underline the specific trajectories of green development practices. In some cases, depending on the complexity and the economic dimension of the contract, the promoters have used characteristic features of GPP, such as: the joint procurement adopted mainly by local governments (small dimension) with the aim to get favorable condition in the tender; the division into sections of smaller technical complexity and cost, therefore, allowing the participation of small and medium enterprises, by not restricting competition; finally, the provision of the pre-commercial phase, useful to investigate on the market, the products or services to be included in the tender, and to assess the needs to be met through the purchase process. It is not clear, within the sample of the case practices analysed, the attention given to social

Figure 6. The environmental impact in the cases examined

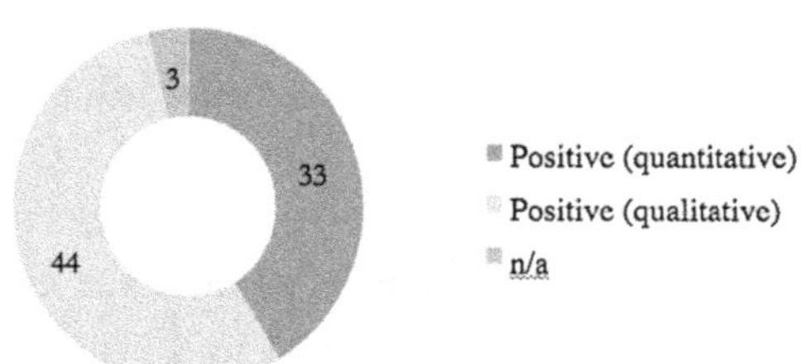

impacts resulting from the purchasing practices, which, however, are most recent actions taken by the European Commission (24 and 25 Directives of 2014).

In addition, worth is noting the possible connection between Tables 1 and 3, therefore, between the existence of the National Action Plan on GPP and the case practices of the countries analysed. Specifically, 71 cases out of 80 are implemented in the Member States, which in line with the European Commission Communications, adopted the NAP's GPP, provided by COM(2003) 302 and 2005 implementation guidelines. In fact, during 2007-2010 the largest number of countries adopted the plans, while a large number of them did so in 2007. In this period were realised most of the case practices examined in the research (Part 3), and recalls the Communications 397 and 400 of 2008 and 400 of 2009. Denmark and Netherlands, in particular, stand out as pioneers in GPP policies implementation, as per 1994 and 2003 plans. The five last cases belongs to countries that do not have prepared any strategy regarding the green public procurement in line with above plan, such as Estonia (with an alternative program), Greece, Romania and Hungary (the plan formulated in 2006-2007 was not adopted). The last four cases, finally, are from Iceland, Norway and Switzerland, which are not part of EU28, and do not have a duty to implement the Communications, nor those concerning NAP's. However, Norway and Switzerland have their environmental policies. On a similar vein, OECD (2009), recognising predominance and importance of the role played by the incubators for sustainable development programs at local government level; in addition, OECD (2013) data shows that these local public authorities exert on average of 55% of procurement spending in OECD countries, with peaks reaching 80% in Spain, Belgium and Italy (Figure 7).

The findings underline that GPP is an effective tool for the rationalization of public spending in the Member States, contributing to the reduction of environmental impacts, energy savings and the promotion of technological innovation. During the process of data gathering of background information, useful for mapping GPP practices, the authors faced the incompleteness of the criteria with which have been selected the best practices cases. Due to such rationales, the analytical framework of the mapped cases does not allow to draw conclusions on the progress of green procurement policies in individual countries. The European Commission, in fact, in promoting the best practices, does not provide aggregate information or evaluation judgments but merely presents the procurement activities, by highlighting the 'excellence' of each procedure in the application of environmental criteria promoted in 'EU. The set of indicators identified throughout the research, laying on the sampled cases examined, aimed to be, in a broader further research agenda, the first step in understanding GPP evolution and practice. The model tends to be replicated to more generic samples, and not necessarily to European Union space.

Figure 7. Share of government contracts by level of government (excluding the social security funds) (2011)
Source: OECD National Account Statistics (database), OECD 2013

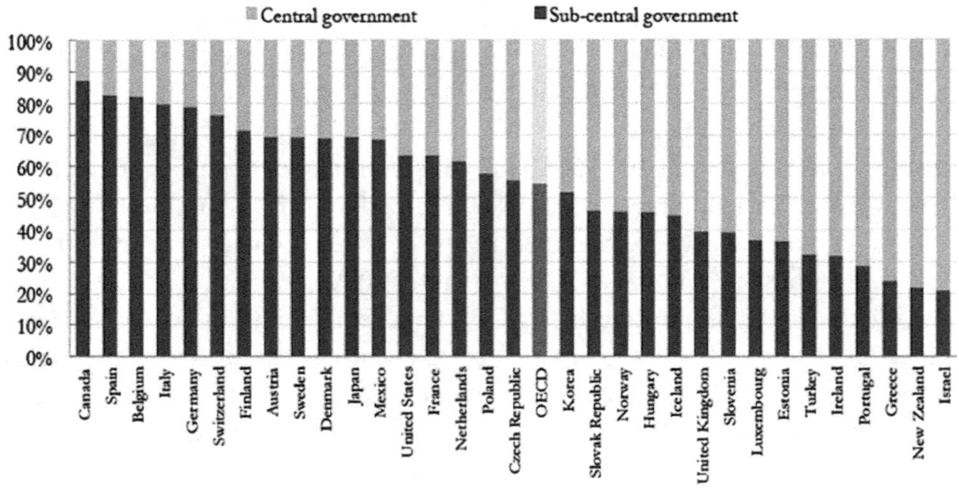

For future research, the authors intend to create a card for the European Commission similar for each case, but more articulated, thus providing a database of information accessible to active public administrations in these practices. By collecting systematic data, based on a simple scheme similar to the current study, alongside with other new useful dimensions to enrich the analysis (including social impacts), it would be possible to track more accurately the trajectory of the European public procurement evolution, alongside a more detailed mapping; also, would encourage the benchmarking and bench learning among the already active public administrations or those interested to be involved in such practices.

REFERENCES

Becchetti, L., Bruni, L., & Zamagni, S. (2010). *Microeconomia*. Bologna: Il Mulino.

Bertoldi, P., Damian Bornás, C., Monni, S., & Piers de Raveschoot, R. (2010). *Linee guida "Come sviluppare un piano di azione per l'energia sostenibile – PAES"*. Unione Europea, Lussemburgo.

Bettini, P., Masini, D., & Ricotta, S. (2006) Acquistare, vendere e rispettare l'ambiente si può, in "Villaggio Globale", n° 36, Edizioni Villaggio Globale, Valenzano (Bari).

Bowen, G. A. (2009). Document Analysis as a Qualitative Research Method. *Qualitative Research Journal*, *9*(2), 27–40.

Carroll, A. B. (1979). A three-dimensional model of corporate social performance. *Academy of Management Review*, 4.

Elkington, J. (1997). *Cannibal with forks: the Triple bottom line of 21st Century Business*. Oxford, UK: Capstone.

European Commission. (1996). Green Paper - Public procurement in the European Union: Exploring the Way Forward. EC.

European Commission. (2001a). Green Paper - Promoting a European framework for Corporate Social Responsibility, COM(2001) 366. EC.

European Commission. (2001b). Communication from the Commission to the Council, the European Parliament, the Economic and Social Committee and the Committee of the Regions, "Environment 2010: Our future, our choice, 6th Environmental Action Programme", COM(2001) 31. EC.

European Commission. (2001c). Communication from the Commission to the Council, the European Parliament, the Economic and Social Committee and the Committee of the Regions, "Interpretative communication of the Commission on the Community law applicable to public procurement and the possibilities for integrating environmental considerations into public procurement", COM(2001) 274. EC.

European Commission. (2003). Communication from the Commission to the Council and the European Parliament, "Integrated Product Policy Building on Environmental Life-Cycle Thinking", COM(2003) 302. EC.

European Commission. (2008a). Communication from the Commission to the Council, the European Parliament, the Economic and Social Committee and the Committee of the Regions, "on the Sustainable Consumption and Production and Sustainable Industrial Policy Action Plan", COM(2008) 397. EC.

European Commission. (2008b). Communication from the Commission to the Council, the European Parliament, the Economic and Social Committee and the Committee of the Regions, "Public procurement for a better environment", COM(2008) 400. EC.

European Commission. (2009). Communication from the Commission to the Council, the European Parliament, the Economic and Social Committee and the Committee of the Regions, "Mainstreaming sustainable development into EU policies: 2009 Review of the European Union Strategy for Sustainable Development", COM(2009) 400. EC.

European Commission. (2010). Communication from the Commission, "Europe 2020. A strategy for smart, sustainable and inclusive growth", COM (2010) 2020 final, 3 March 2010. EC.

European Commission. (2011a). Direttiva del Parlamento Europeo e del Consiglio on Public Procurement, COM(2011) 896. EC.

European Commission. (2011b). *Commission staff working paper, Evaluation Report, "Impact and Effectiveness of EU Public Procurement Legislation", SEC(2011) 853*. EC.

European Commission. (2011c). Communication from the Commission to the Council, the European Parliament, the Economic and Social Committee and the Committee of the Regions, "Energy Roadmap 2050", COM(2011) 885. EC.

European Commission. (2011d). Buying Green! A Handbook on Green Public Procurement. EC.

European Commission. (2012). *Green Public Procurement, A collection of good practices*. EC.

European Commission. (2014a). Retrieved from http://ec.europa.eu/environment/gpp

European Commission. (2014b). Retrieved from http://www.europarl.europa.eu

European Commission. (2004a). Directive 2004/17/EC of the European Parliament and of the Council of 31 March 2004, "coordinating the procurement procedures of entities operating in the water, energy, transport and postal services sector" (2004/17/EC). *Official Journal of the European Union*, 1-113.

European Commission. (2004b). Directive 2004/18/EC of the European Parliament and of the Council of 31 March 2004, "coordinating the procedures for the award of public works contracts, public supply contracts and public service contracts" (2004/18/EC). *Official Journal of the European Union*, 114-240.

European Commission. (2014c). Directive 2014/23/EU of the European Parliament and of the Council of 26 February 2014 on the award of concession contracts. *Official Journal of the European Union (28/03/2014)*, 1-64.

European Commission. (2014d). Directive 2014/24/EU of the European Parliament and of the Council of 26 February 2014 on public procurement and repealing Directive 2004/18/EC. *Official Journal of the European Union (28/03/2014)*, 1-178.

European Commission. (2014e). Directive 2014/25/EU of the European Parliament and of the Council of 26 February 2014 on procurement by entities operating in the water, energy, transport and postal services sectors and repealing Directive 2004/17/EC. *Official Journal of the European Union (28/03/2014)*, 1-132.

Fiorani, G., Jannelli, R., & Meneguzzo, M. (2012). *CSR 2.0 Proattiva e sostenibile*. Milano: Egea.

MATTM. (2013). Piano d'Azione per la sostenibilità ambientale dei consumi nel settore della Pubblica Amministrazione – Revisione 2013. *Gazzetta Ufficiale della Repubblica Italiana, 102*, 17–29.

Meneguzzo. (2012). *Managerialità innovazione e governance*. Archidata.

OECD. (1996). *Ministerial Symposium on the future of public services*. OECD/PUMA.

OECD. (2009). *Cities, Climate Change and Multilevel Governance*. OECD Environmental Working Papers n°14, 2009. OECD Publishing.

OECD. (2013). *Government at a Glance 2013*. OECD Publishing.

ONU. (2013). Retrieved from http://unstats.un.org/unsd/methods/m49/m49regin.htm#europe

PricewaterhouseCoopers. (2009). *Collection of statistical information on Green Public Procurement in the EU*. Author.

Ricotta, S. (2012). Il Green Public Procurement per il cambiamento climatico. Gazzetta Ambiente, 2.

Saltari, E., & Travaglini, G. (2005). *Le radici del declino economico*. Utet Università.

Tukker, Emmert, S., Charter, M., Vezzoli, C., Sto, E., Munch Andersen, M., ... Lahlou, S. (2008). Fostering change to sustainable consumption and production: An evidence based view. *Journal of Cleaner Production*, *16*(11), 1218–1225. doi:10.1016/j.jclepro.2007.08.015

UNCED. (1992). Agenda 21. *United Nations Conference on Environment & Development*, Rio de Janeiro, Brazil.

Zuppinger, P. (2005). *Green Public procurement in Europe 2005 - Status Overview*. Haarlem: Virage Milieu & Management.

Chapter 5
A Perception on Green Production in Textile Industry for Environmental Sustainability:
Detection Strategy With SWOT and AHP Analysis

Ayşenur Erdil
Yalova-Marmara University, Turkey

Mehtap Erdil
Marmara Istanbul University, Turkey

ABSTRACT

The purpose of this chapter is to investigate and determine the most appropriate strategy according to the results of applied methodologies and literature about waste management for the textile sector. In this research, improvement of this concept for textile sector in Turkey has been implemented by using SWOT analysis integrated with AHP framework. Based on this, the most appropriate strategy has been chosen for this industry. According to this destination, SWOT analysis has been applied and alternative strategies have been determined which are based on green production and waste management of textile products for this sector. Then this SWOT matrix has been converted into a hierarchical structure. Thus, AHP model of this application has been formed with this perception via this methodology. Consequently, determination a management strategy within environmentally oriented production for this industry will be provided to contribute the environmental sustainability. Also, this study will be given a point of general view and approach for the importance of green production and consumption.

DOI: 10.4018/978-1-5225-7083-7.ch005

INTRODUCTION

Industrial Environment such as a content involves usage of a symbol (implied comparison) complement industrial frameworks with essential ecological life (Bouten et al, 2011). Natural ecological life operations interconnect and determine in a method to guarantee the continuation of this structure. So, this frame must support to provide a method to optimize capability so as to alive and sustain from the probable reduction of natural and financial sources (Bcst, 2005).

Possible for green development according to production framework of the system is suitable for the ecological area and goods safety procedures. Although, shortening life cycle of products consume harmful matters for generating inputs need more suitable waste management. For instance, much more amount of electric and electronic disposal, garbage consisted of valued matters like copper and gold extra harmful and dangerous elements were composed and converted.

Some of the significant issues according to this scope are listed as below:

- **Environmental and Social Impacts of Clothing:** "The ecological influences of apparel along the supply chain of clothing contain these issues:
    - Energy consumption and producing and generating of Greenhouse Gas (GhG) diffusions from the actions of washing-cleaning via the high temperatures"water-heating and drying of"garment;
    - Energy consumption, resource reduction and producing of GhG diffusions-footprints from dealing with fossil fuels with synthetic threads for instance polyester and nylon;
    - Consuming important and essential amount of water, hazardous toxicity, chemicals from ammonium nitrate, fertilizer, using chemical substance and weedkiller, consuming energy and GhG footprints-diffusions depending with fertilizer-potassium nitrate generation and irrigation structures outside of fibre crops (Maxwell, 2008; Defra, 2008; Turkey's Sustainable Development Report, 2012; OECD, 2001).
- **Sustainable Apparel-Garment:** It is defined that Apparel which supports to maximize-increase the positive effects and minimize-decrease the negative impacts of environmental-ecological, social and financial-economic influences towards this supply chain structure and value chain of this frame. Apparel which is sustainable, may not negatively affect householders or the world in this manufacturing-generating, transport, sale, marketing or the end of life cycle time arrangement. In an implementation, providing and supporting this frame is not uninterrupted progressive, includes bargain, arrangement and the commercial exchange (buying and selling on domestic or international markets) of goods and services "between different influences and planned

out developments accomplished the types of periods (short-lived, average and lengthy term). Declaring and the activity of conveying information "sustainable" apparel in periods which whole shareholders, particularly customers show a response or a reaction to something is a significant point condition in the schedule (Defra, 2008). By implementing activities in these scopes together for business and customer coming up against shareholders-collaborators can develop the sustainability of garments. This structure provides changing attitudes, habits of consumers to diminish influences of garments buying, repairing and the amount of waste-disposal. Some of them are listed below (Defra, 2008; Turkey's Sustainable Development Report, 2012; OECD, 2001):

- ◦ Improving and suggesting varieties of the garment which have developed the dimensions' qualities of environmental and social impacts.
- ◦ Giving information and providing assistance for consumers in scopes where they could want to change their apparel.
- ◦ Repairing-preservation apparel for the minimum energy consumption and chemical substance accelerated method.
- ◦ Using again and reprocessing, using again after processing of undesired apparel-garments.

- **Developing Ecological-Environmental Accomplishment Crossways the Supply Chain:** Invariable garment structure for designing and improving the apparel products which includes the effects of the garment on the ecological area by measuring the volume of disposal-waste generated during the design and producing states, consuming environmentally, selected materials and by removing the use of chemicals-toxins. Sustainable Matters-Substances: Developing the technique for determining ecological influences and evaluating materials for materials aggregation-collection (Maxwell, 2008).

IMPROVING TRACEABILITY WITHIN SUPPLY CHAIN STRUCTURE

Forming, Organizing and sustaining a"fair and reliable-quality third party verification structure is the application of the cotton, wool"produced with Sustainability Principles-Regulations which relates and provides social, economic-financial, environmental properties. There must be equality and reliability between environmental-ecological life, commerce business and ethics features in this Supply Chain System.

When an apparel product-textile material has been produced and this possible for purposing harmful to the ecological life does not finish. A textile cloth has a lot of production phases. The potential goods of this next processing involve much more extensive components such as apparel, garment curtain, bedding and various industrial products. Business generally attends to relate the basic (for instance fabric generation) and subordinate (for instance goods produced from textile materials) apparel-clothing sectors like disconnected ones, however, the ecological life causes any similar differentiation. One of the significant issues causing the integrated industry currently is a deficiency of connection among these two various fields, because this situation happens the lack of misunderstanding of the capabilities of the basic-main industry and the lack of misconception of the requirements of the secondary-subordinate industry. The general scope might provide considerably by the improvement of nearer relation among the two sector sides, decreasing a number of discarded-disposal source materials which conclude due to these defects to integrate requirements, capabilities, and generation (Defra, 2008; Slater, 2003).

The scope of ecological protective covering in the apparel-clothing concept involves much more general goods depending on the requirement for the business sector to find solution methods for pollution issues. Germany confirms-launches strategic policy which includes regulations to restrict or exclude ecological and health endangerment-risks for apparel-clothing and stages of leather production industries. Japanese clothing business focuses on harmonization with the ecological area, noticing for instance and the potential of reusing the used garments. These also meet industrial customers to the opinions of generating environmentally, ecologically friendly goods, saving and consuming the lowest amount of energy sources, and supporting to decrease water contamination – deterioration (Slater, 2003).

A periodic time of ecological key point is focused by Slater who searches and thinks that this situation could have risk potentials for some business sectors according to the ecological and environmental objection. He defines the requirements for more security and better analysis for decreasing contamination to provide the recently integrating regulations. Shaver emphasizes the requirement for firms to provide their organization of environmental management accountabilities due to the organized increment in expenditures and responsibilities. He advises an ecological management structure as an effective method to organize-build degree of engagement, improving plans, schedules or controls and adjusting issues. By its endorsement, the ecological influences of production-manufacturing will happen accessibly and a decreasing for expenditures would be attended by an income in the introduction for the business (Slater, 2003; Shaver, 1999).

SWOT ANALYSIS CONCEPT

The management draws a general picture of the environment around the organization from raw information gathered about the organization itself, the industry, and the macro environments. However, environmental factors must be assessed systematically to be able to interpret the organization's environments and understand what they mean for the organization. It is possible to understand and manage the organization's environment thusly. SWOT analysis is a way to achieve this (Dinçer, 2004; Ülgen and Mirze, 2004). In this regard, SWOT analysis can be defined as an organization's systematic self-assessment using information it has gathered to understand and manage the environment in which it operates. These assessments are often influenced by personal judgments and understanding (Ülgen ve Mirze, 2004; Dinçer, 2004). As an example, the lowering of customs walls can be considered as a threat and still as an opportunity by some other managers. The SWOT analysis provides the management with systematic information about the organization's the environment. How administrators interpret, this information is related to their management understanding (Kajanus et al., 2004; Kurttila et al., 2000; Leskinen et al., 2004; Masozera et al., 2004; Shresta et al., 2004).

SWOT stands for advantages, weaknesses, opportunities, and threats.

With SWOT analysis, an organization can evaluate its strengths and weaknesses within itself as well as opportunities and threats originating from its external environment. This way it can determine the most significant factors surrounding the organization (Hill and Westbrook, 1997). A good strategy means that a firm ensures harmony between its internal qualities or characteristics (strengths and weaknesses) and the external conditions (opportunities and threats) it encounters. Thus, the strategy choice takes place in accordance with the decision maker's targets(Kajanus et al., 2004). SWOT analysis means analyzing critical or key success factors, and it should be considered as an important step in developing a good strategy(Bernroider, 2002: 564). When used correctly, the SWOT analysis can provide a good basis for strategy formulation (Kajanus et al., 2004). The SWOT analysis can be said to have certain shortcomings despite its common application area. One of the most important limitations of the SWOT analysis is the inability to quantify the importance of each factor in decision making. Moreover, it is difficult to evaluate through this analysis which factor has the greatest impact on the strategic decision (Shrestha et al., 2004).

THE PURPOSE OF SWOT ANALYSIS

SWOT analysis enables managers to understand the factors that significantly affect the firm's performance. Also, it helps them determine how to take action against them. These factors are shortly called strategic issues. A strategic issue is any factor involved in the organization affecting its accomplishment of its goals. It is difficult to ensure that strategic issues reach the top management's desk after being fully sorted (Şeker and Özgürler, 2012). Generally, the data obtained after the SWOT come to management's desk as new technologies, new market trends, new competitors, and employees' motivation. They need to be interpreted and transformed into a strategic management language to become a strategy. Depending on the management's knowledge, experience, and intuitions, these subjects are classified as controllable/ uncontrollable, strength, weakness, opportunity, and threat. Then, managers exchange opinions. They state how they assess the strategic topic and decide on the required initiatives. It is clear that every strategic issue does not have equal importance for every organization. Some organizations are more sensitive to certain issues than others. It is a major challenge for senior management to determine which matter is more important for the organization than others, and a determinant of success. Different managers put forward different opinions (Lee et al., 2011). SWOT analysis reveals the business's weakness and strengths as well as possible threats and opportunities for the business. The aim is to ensure that the business has maximum benefit from opportunities and puts forward strengths by taking into account the weaknesses and threats (Stewart et al., 2005). Therefore; It presents existing and future threats while taking counter measures. The weaknesses of the business will be determined and improved Strategies and action plans will also be developed to maximize the benefits from opportunities by bringing the strengths of the business to the forefront (Masozera et al., 2004).

APPLIED METHODS AND WORKING PROCEDURE

Participants to conduct the SWOT analysis are employees that have the knowledge, skills, and formation to represent the firm as well as individuals enjoying expertise in business administration activities. Despite various methods, the brainstorming technique that the analysis team (Participants) performs together is the most preferred one to obtain healthy results (Stewart et al., 2005; Dyson, 2004).

The steps of the proposed working procedure for SWOT analysis are as follows (Harrison, 2010):

- Interviews with relevant experts with company officials and determination of the scope of the SWOT analysis.
- Creation of working team for the specified scope
- Collection of information through SWOT,
- Working team's detailed examination of the production process,
- Having the firm meeting to present the firm's strengths and weaknesses, opportunities, and threats,
- The assessment of the identified strengths and weaknesses by the working team and completion of the missing parts,
- Making an evaluation meeting with the participation of company officials and working team,
- Interpretation of SWOT analysis, determination of strategies and presentation of the action plan.

Thus, this study, which takes about 23 hours based on the time spent in the firm, will be completed in 4-5 weeks on average.

APPLIED METHODOLOGY: ANALYTIC HIERARCHY PROCESS

Developed by Saaty in 1980, the AHP is a useful decision support tool for managing multi-criteria decisions. AHP is an effective tool for structuring and modeling multi-criteria problems. It has been successfully used in the structuring of various management applications. The AHP divides the problem into pieces and eventually combines all the solutions related to these parts. AHP shows all the forces that influence a decision by organizing intuitions, emotions, judgment, and intelligence together and facilitates decision-making (Saaty, 1980).

The problem is structured in a hierarchical manner in AHP. The prioritization process follows the hierarchical structuring of the problem such as Figure 1.

The AHP's second fundamental step is binary comparisons. A binary comparison is the comparison of two factors or criteria, and it is based on a decision maker's judgment. With binary comparison, the goal is to determine the relative importance of the elements in the hierarchy relative to an element at a higher level. Therefore, a binary comparison is designed for the establishment of priority distributions of decision criteria and alternatives. When there are n number of elements at the level of the hierarchy to be compared, it is necessary to perform n (n-1)/2 comparisons. Each comparison should be arranged in matrix form. Saaty developed the 1-9 scale to assign values in binary comparison as seen in Table 1. This scale is generally used in AHP (Saaty, 1980).

Figure 1. Three-level AHP model

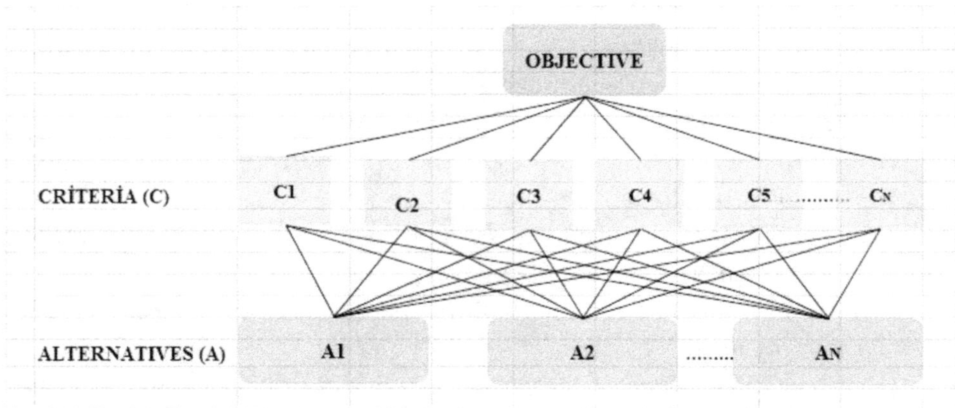

Table 1. Relative importance scale

Importance Level	Definition	Explanation
1	Equal importance	Two activities contribute equally to the objective
3	Moderate importance	Experience and judgement slightly favour one activity over another
5	Strong importance	Experience and judgement strongly favour one activity over another
7	Very strong or demonstrated importance	An activity is favoured very strongly over another; its dominance demonstrated in practice
9	Extreme importance	The evidence favouring one activity over another is of the highest possible order of affirmation
2,4,6,8	Intermediate values	It is used when you need to give a middle value between two options.

(Modified by Author from Thomas L. Saaty & Mariya Sodenkamp, 2008)

The matrix A to be created for a goal with n criteria has a size of n x n. The binary comparison matrix is constructed as in Table 2.

In the binary comparison matrix, there are values indicating the importance of the row i element with respect to the column j element. Wi/wj on the table is a term that expresses the importance of criterion i. over criterion j. The binary comparison matrix needs to be normalized after its preparation. To normalize the matrix, one should take the sum of columns for each column, and divide the matrix elements by the sum of the relevant columns. Then, one should take the sum of rows for each alternative or criterion in the normalized matrix. The resulting values are the priority

Table 2. Binary comparison matrix

$$A= \begin{matrix} A_1 \\ \cdot \\ \cdot \\ A_n \end{matrix} \begin{pmatrix} A_1 & \cdot & \cdot & \cdot & A_n \\ w_1/w_1 & \cdot & \cdot & \cdot & w_1/w_n \\ \cdot & \cdot & \cdot & \cdot & \cdot \\ \cdot & \cdot & \cdot & \cdot & \cdot \\ w_n/w_1 & \cdot & \cdot & \cdot & w_n/w_n \end{pmatrix}$$

values for the criteria or alternatives. The matrix formed by the priority values is the priority vector matrix. The priority value for each criterion/alternative in the priority vector matrix is multiplied by all the elements in the column of such criterion's/alternative's binary comparison matrix to obtain the weighted total matrix. The total values of the rows within the weighted total matrix are divided by the row values of the priority vector matrix. The arithmetic mean of the values in the last matrix enables the calculation of the eigenvalue. After these steps, the Consistency Index and Consistency Rate should be calculated. The following formulas are used in the calculation of the consistency ratio of matrix A:

$$CR= CI / RI$$

$$CI= (\lambda_{max}-n) / (n-1)$$

CI: Consistency Index
RI: Random Index
CR: Consistency Ratio

Generally, the matrix is considered to be consistent if the consistency ratio (CR) is 10% or less. Also, if the largest eigenvalue is equal to the matrix size ($\lambda max = n$), the comparison matrix is said to be consistent (Saaty, 1991).

COEXISTENCE OF SWOT AND AHP IN DETERMINING STRATEGY

AHP is considered to be the most appropriate analytical method for developing a hybrid method with SWOT. The use of AHP with SWOT provides analytically determined priorities for the factors involved in SWOT analysis and ensures proportionality of such factors. Addressing the AHP and the SWOT analyses are better than conducting a SWOT analysis per se. SWOT analysis does not explain the importance of factors to determine the impact of each factor on the presented plan or strategy. The outcome of the SWOT analysis is mostly a list of internal and external environmental factors or an incomplete qualitative study. However, the combination of the two techniques allows one to know the strengths, weaknesses, opportunities, and intensities of the threats or the most important group based on quantitative AHP prioritization. The most important group can not be fully identified by focusing solely on the weighting of SWOT factors. AHP allows decision makers to assign a relative priority to each factor by making a bilateral comparison. One of the possible advantages of using AHP in SWOT analysis is the quantitative examination of SWOT factors and including decision-makers' choices. The decision maker will have new quantitative information about the situation that will be decided after addressing AHP and SWOT analysis together (Shrestha et al., 2004; Duru et al., 2013).

The literature presents many studies in which the AHP and ANP methods are applied with fuzzy numbers. For example, whereas Lee et al. (2008) brought the fuzzy AHP together with balanced goal-card methods in performance evaluation, Pan (2008) used the fuzzy AHP method in the selection of the most appropriate bridge construction method. While Vahindia and others(2009) searched for a solution to the hospital location selection problem with the fuzzy AHP method with Cebeci also benefiting from the same method in selecting corporate resource planning. On the other hand, Yüksel and Dağdeviren (2010) suggested using a fuzzy ANP method and balanced goal cards when evaluating a production company's performance. Chen and Chen (2010) used the fuzzy ANP method together with other evaluation methods in decision making. Also, Promentilla et al. (2008) reached a conclusion with the fuzzy ANP method in the multi-criteria evaluation. The literature reveals studies where AHP and ANP methods are used together although not many. Examples of these studies include Yang et al.'s (2009) use of AHP/ANP methods in the evaluation system of production, Saaty's (2007) application of the AHP/ANP methods in time-dependent decision-making, and Garuti and Spencer's (2007) examination of the alignment between these two methods. The study offering a foundation for our study is the one conducted by Kahraman and others (2007) using SWOT analysis and AHP method together in the prioritization of e-government strategies. There is an effort to integrate the SWOT analysis, which has been an important research topic for many years in the literature, with AHP and ANP methods, the multi-criteria selection methods used in recent years. At this point, while Yüksel and Dağdeviren (2007) use the SWOT analysis and ANP method in the textile industry, Kurttilla et al. integrated the SWOT analysis and AHP method in strategic planning. Shrestha et al. (2004) used the same methods for application in the agricultural industry.

SWOT ANALYSIS APPLICATION

This study basically followed the steps taken by Yüksel and Akın (2006) in determining strategy while using AHS and SWOT techniques together. Therefore, we first conducted an internal and external environmental analysis to determine the SWOT factors which could affect success or failure. We had interviews with experts in the textile industry to determine the degree of importance of the SWOT group factors of the faculty applying Likert-type questionnaires. We carried out assessments by taking expert opinions. We held industry-focused meetings. We contacted the experts from the chambers of commerce operating in all regions and the Turkish Exporters Assembly to reflect the textile industry across Turkey. We took their opinions and recommendations, too. We obtained consumer opinions were through a short survey on the textile industry's consumer segment. In the questionnaire, we indicated the strength, weakness, opportunity and threat elements for the industry and asked their importance level. The averages of the data obtained from the questionnaire were taken using IBM SPSS package program. Relative strengths were determined by taking into account these comparisons during binary comparison of the factors. This point forms one of the unique aspects of the study. When making binary comparisons of SWOT group factors, the relative strengths were determined by taking into account the average responses of the experts and conscious consumers (customers) to the questionnaire. Therefore, the figures in the binary comparison matrices were given by persons knowing the industry closely. During the application of the method, the values in the binary comparison matrices were determined based on the scale developed by Saaty (Table 1) The model was analyzed on the computer using Expert Choice 11 software.

FINDINGS AND RESULTS

The overall draft Swot Matrix for the textile industry in Strategy Selection using Analytic Hierarchy Process is as in Table 3.

Table 3. The draft swot matrix framework

	POSITIVE	NEGATIVE
INTERNAL	STRENGTHS (ST)	WEAKNESSES (W)
EXTERNAL	OPPORTUNITIES (O)	THREATS (T)

As a result of the inspection, Factors in the Strength group of Textile-Garment Industry were specified as

- Green Design; Decrease greenhouse gases (GHGs) (ST1)
- Strong Research and Development for steps of process of production--Relations in the sustainable of green supply chain can provide to practice green implementation (ST2)
- Improve alternatives for insufficient appropriate resources (energy, materials); Decrease water consumption of manufacturing process (ST3) and
- Upgrade corporate image and product reputation, increase market share ST4)
- The Weaknesses Factors of the sector were declared as
- Lack of technical experts and experience; Lack of ecological knowledge (W1)
- Effect of Historical Government Policies (W2)
- Technological Obsolescence; Expenditure of environmentally friendly packaging (W3) and
- Requirement Research and Development expenditure to determine green solutions for adapting technology (W4)
- On the other hand, the Factors that create Opportunities for the sustainability of this sector were discovered as,
- Maximize reuse and minimize waste (O1)
- Provide advantage of government incentives for manufacturing green products (O2)
- Environmental challenges requires innovation and collective action by multiple parties across the value chain and beyond (O3) and
- Support to increase demand of customer for green products/services (O4).
- Factors that constitute Threats for the sector were determined as,

- Lack of participation of top management in acquiring green business, green production (T1)
- Lack of global green and quality certification system can support pretended green products (T2)
- Increasing the carbon, water and energy footprints (T3) and
- Competition in Domestic Market (Market competition) (T4).

After determining the SWOT factors, alternative strategies have been developed according to these above factors.

When we look at the formation of the General Strategy, the following structure is seen such as Table 4.

As can be seen from the SWOT matrix, the industry's alternative strategies include these items:

- Developing systems to optimize material transportation within the factory,
- Working to reduce fuel consumption in product/service deliveries, and
- The green workforce that can apply technology is specified as S1;
- Specifying and determining alternatives rather than toxic chemicals,
- Taking measures to reduce greenhouse gas emissions,

Table 4. SWOT matrix of textile industry

STRATEGIES		
	STRENGTHS	*WEAKNESSES*
THREATS	• Assuming that consumers will prefer green products and paying a high price for them if necessary, • Using eco-labeling and achieving competitive advantage, • Positive effect of environment-oriented measures on competitiveness, • The application of the ISO 14001 environmental management standard, The management's support and reward for practices that lead to green change. (S3) (MAXI-MINI)	• Developing systems to optimize material transportation within the factory, • Working to reduce fuel consumption in product/service deliveries, • The green workforce that can apply technology. (S1) (MINI-MINI)
OPPORTUNITIES	• Reducing costs through resource efficiency, • Reducing costs by decreasing the waste amount, • Ensuring that green, clean, environmentalist product development provides competitive advantage; • Developing and executing activities for the design of green-focused products, • Taking package-reducing and recycling measures in design, • A sustainable, recyclable and renewable material has been selected as a production input. (S4) (MAXI-MAXI)	• Specifying and determining alternatives rather than toxic chemicals, • Taking measures to reduce greenhouse gas emissions, • Reducing water consumption and saving water, • Setting a target to reduce energy use. (S2) (MINI-MAX))

- Reducing water consumption and saving water, and
- Setting a target to reduce energy use, are specified as S2
- Assuming that consumers will prefer green products and pay a high price for them if necessary,
- Using eco-labeling and achieving competitive advantage,
- Positive effect of environment-oriented measures on competitiveness, and
- The application of the ISO 14001 environmental management standard, the management's support and reward for practices that lead to green change, are specified as S3
- Reducing costs through resource efficiency,
- Reducing costs by decreasing the waste amount,
- Ensuring that green, clean, environmentalist product development provides competitive advantage;
- Developing and executing activities for the design of green-focused products,
- Taking package-reducing and recycling measures in design, and
- A sustainable, recyclable and renewable material has been selected as a production input, are specified as S4.

The problem has been hierarchically structured in four levels to be able to quantify the set of strategic factors and alternative strategies identified as a result of the SWOT analysis using the AHP technique. The problem has been hierarchically structured in four levels to be able to quantify the set of strategic factors and alternative strategies identified as a result of the SWOT analysis using the AHS technique. The hierarchical state of the problem can be seen in Figure 2. The model's first, second, third and fourth levels include the aim of choosing the most appropriate strategy; SWOT groups; sub-factors constituting the SWOT groups, and the four alternative strategies identified, respectively.

The dual comparison matrix of SWOT group factors and local and general weights of the factors are shown in Table 5.

According to Table 4, while opportunities are more important than threats, strengths do not matter as much as opportunities. Likewise, weaknesses are not as important as opportunities. Strengths are more important than threats, and weaknesses are more important than threats. Weaknesses have more importance than strengths. The binary comparison matrices of the sub-factors that constitute the SWOT group at the third level of the AHP model are given in Annex 1. On the other hand, Annex 1, Annex 2, Annex 3, Annex 4, and Annex 5 provide the binary comparison matrices of the strategies within the model's fourth level for the opportunities, threats, strengths, and weaknesses, respectively. The local and general weights obtained from the resolution of the model are shown collectively in Table 5. The local weights provided in the table were found at the end of the binary comparison

Figure 2. Analytic Hierarchy Process Model of the Problem: Starting Model

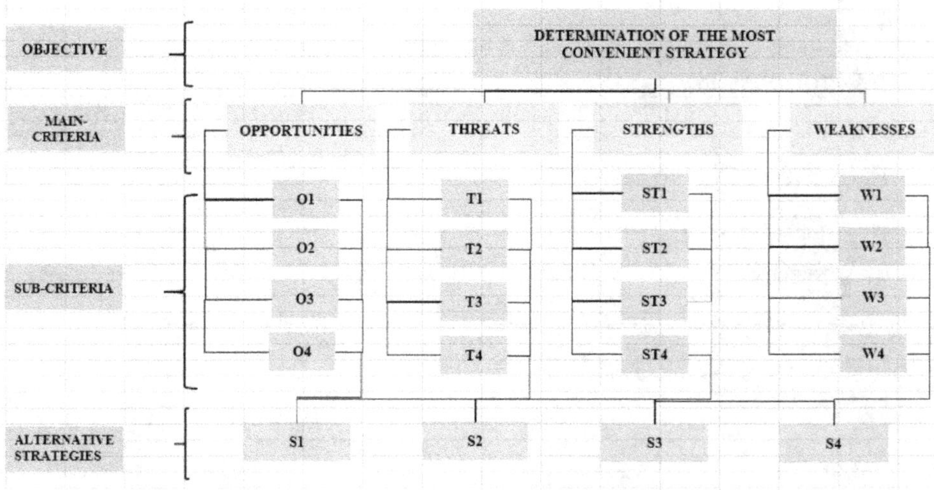

for each factor or strategy, and the general weights are the values representing the shares of the SWOT group factors in the total weight. These values, i.e. local and general weights, were determined by the Expert Choice program after the solution of the problem. According to the tables in the Appendix, the local weights of the sub-factors in the Opportunities group are as follows.

According to the table in the Appendix, the local weights of the sub-factors in the Opportunities group are as follows: It was calculated as 19.1% for O1- Maximize reuse and minimize waste; It was calculated as 30% for O2- Provide advantage of government incentives; It was calculated as 19.7% for O3- Environmental challenges requires innovation and collective action by multiple parties across the value chain and beyond; It was calculated as 31.2% for O4- Support to increase demand of customer for green products/services. In this case, the most important subfactor in the opportunity group is O4- Support to increase demand of customer for green products/services.

According to the table in the Appendix, the local weights of the sub-factors in the Threats group are as follows. 37.1% for T1- Maximize reuse and minimize waste; It was calculated as 36.5% for T2- Lack of global green and quality certification system can support pretended green products; It was calculated as 13.8% for, T3- Increasing the carbon, water and energy footprints; It was calculated as 12.6% for T4- Competition in Domestic Market (Market competition) In this case, the most important sub-factor in the opportunity group is T1- Maximize reuse and minimize waste.

Table 5. Normalization of main criteria: Dual comparison matrix of SWOT group factors

COMPARISON OF MAIN CRITERIA				
	STRENGTHS	WEAKNESSES	OPPORTUNITIES	THREATS
STRENGTHS	1.0000	0.3333	0.2500	2.8570
WEAKNESSES	3.0000	1.0000	0.2000	2.3255
OPPORTUNITIES	4.0000	5.0000	1.0000	2.8570
THREATS	0.3500	0.4300	0.3500	1.0000
TOTAL	8.3500	6.7633	1.8000	9.0395

NORMALIZATION OF MAIN CRITERIA					
	STRENGTHS	WEAKNESSES	OPPORTUNITIES	THREATS	WEIGHT
STRENGTHS	0.1198	0.0493	0.1389	0.3161	0.1560
WEAKNESSES	0.3593	0.1479	0.1111	0.2573	0.2189
OPPORTUNITIES	0.4790	0.7393	0.5556	0.3161	0.5225
THREATS	0.0419	0.0636	0.1944	0.1106	0.1026

According to the table in the Appendix, the local weights of the sub-factors in the Strengths group are as follows. It was calculated as 15.4% for S1- Maximize reuse and minimize waste; It was calculated as 12.2% for S2- Strong Research and Development for steps of process of production, Relations in the sustainable of green supply chain can provide to practice green implementation; It was calculated as 11.6% for S3-Improve alternatives for insufficient appropriate resources (energy, materials); Decrease water consumption of manufacturing process ; It was calculated as 60.8% for S4- Upgrade corporate image and product reputation, increase market share. In this case, the most important sub-factor in the opportunity group is S4- Upgrade corporate image and product reputation, increase market share.

According to the table in the Appendix, the local weights of the sub-factors in the Weaknesses group are as follows. It was calculated as 7.8% for W1- Lack of technical experts and experience; Lack of ecological knowledge; It was calculated as 11.7% for W2- Effect of Historical Government Policies; It was calculated as 51.1% for W3- Technological Obsolescence; Expenditure of environmentally friendly packaging; It was calculated as 29.4% for W4- Requirement Research and Development expenditure to determine green solutions for adapting technology. In this case, the most important sub-factor in the opportunity group is W3-Technological Obsolescence; Expenditure of environmentally friendly packaging.

As can be seen from Table 6, the overall weights of each sub-factor that constituted the SWOT groups were also calculated.

General Weights of All Sub factors value were sorted as below:

- General weight of sub-factor of Strengths Main Criteria was calculated as 2.4% for (ST1) Green Design; Decrease greenhouse gases (GHGs);It was calculated as 1.9% for (ST2) Strong Research and Development for steps of process of production--Relations in the sustainable of green supply chain can provide to practice green implementation; It was calculated as 1.8% for (ST3) Improve alternatives for insufficient appropriate resources (energy, materials), Decrease water consumption of manufacturing process; It was calculated as 9.4% for (ST4) Upgrade corporate image and product reputation, increase market share;
- General weight of sub-factor of Weaknesses Main Criteria was calculated as 1.7% for (W1) Lack of technical experts and experience, Lack of ecological knowledge; It was calculated as 2.5% for (W2) Effect of Historical Government Policies; It was calculated as 11.1% for (W3) Technological Obsolescence, Expenditure of environmentally friendly packaging; It was calculated as 6.4% for (W4) Requirement Research and Development expenditure to determine green solutions for adapting technology;
- General weight of sub-factor of Opportunities Main Criteria was calculated as 9.9% for (O1) Maximize reuse and minimize waste; It was calculated as 15.6% for (O2) Provide advantage of government incentives for manufacturing green products; It was calculated as 10.2% for (O3) Environmental challenges requires innovation and collective action by multiple parties across the value chain and beyond; It was calculated as 16.2% for (O4) Support to increase demand of customer for green products/services;

Table 6. Local and general weights for binary comparisons of basic and sub factors

MAIN CRITERIA	WEIGHTS	SUB CRITERIA	WEIGHTS	GENERALWEIGHTS	STRATEGIES	WEIGHTS
STRENGTHS(ST)	0,156	ST1	0,1537	0,0240	S1	0,110
					S2	0,105
					S3	0,382
					S4	0,404
		ST2	0,1222	0,0191	S1	0,102
					S2	0,325
					S3	0,439
					S4	0,134
		ST3	0,1160	0,0181	S1	0,372
					S2	0,115
					S3	0,077
					S4	0,436
		ST4	0,6082	0,0949	S1	0,383
					S2	0,342
					S3	0,186
					S4	0,089
WEAKNESSES	0,218	W1	0,0783	0,0171	S1	0,401
					S2	0,370
					S3	0,077
					S4	0,153
		W2	0,1171	0,0255	S1	0,206
					S2	0,176
					S3	0,497
					S4	0,120
		W3	0,5108	0,1114	S1	0,095
					S2	0,552
					S3	0,200
					S4	0,152
		W4	0,2938	0,0640	S1	0,437
					S2	0,386
					S3	0,078
					S4	0,099
MAIN CRITERIA	WEIGHTS	SUB CRITERIA	WEIGHTS	GENERALWEIGHTS	STRATEGIES	WEIGHTS
OPPORTUNITIES	0,522	O1	0,1909	0,0997	S1	0,206
					S2	0,176
					S3	0,497
					S4	0,120
		O2	0,3005	0,1568	S1	0,144
					S2	0,070
					S3	0,255
					S4	0,531
		O3	0,1967	0,1027	S1	0,132
					S2	0,552
					S3	0,200
					S4	0,152
		O4	0,3119	0,1628	S1	0,166
					S2	0,632
					S3	0,115
					S4	0,087
THREATS	0,103	T1	0,4273	0,0440	S1	0,049
					S2	0,623
					S3	0,206
					S4	0,122
		T2	0,3258	0,0336	S1	0,527
					S2	0,110
					S3	0,258
					S4	0,105
		T3	0,1368	0,0141	S1	0,078
					S2	0,184
					S3	0,475
					S4	0,263
		T4	0,1102	0,0114	S1	0,068
					S2	0,127
					S3	0,598
					S4	0,207

- General weight of sub-factor of Threats Main Criteria was calculated as 4.4.% for (T1) Lack of participation of top management in acquiring green business, green production; It was calculated as 3.36% for (T2) Lack of global green and quality certification system can support pretended green products; It was calculated as 1.41% for (T3) Increasing the carbon, water and energy footprints (T3); It was calculated as 1.14% for (T3) Competition in Domestic Market (Market competition).

Followings are the strategies regarding the Textile Industry's environmental sustainability:

- Developing systems to optimize material transportation within the factory, working to reduce fuel consumption in product/service deliveries, the green workforce that can apply technology are determined as S1(the first strategy),
- Specifying and determining alternatives rather than toxic chemicals, taking measures to reduce greenhouse gas emissions, reducing water consumption and saving water, setting a target to reduce energy use are determined as S2(the second strategy),
- Assuming that consumers will prefer green products and pay a high price for them if necessary; using eco-labeling and achieving competitive advantage; positive effect of environment-oriented measures on competitiveness; the application of the ISO 14001 environmental management standard; and the management's support and reward for practices that lead to green change are determined as S3 (the third strategy),
- Reducing costs through resource efficiency, reducing costs by decreasing the waste amount, ensuring that green, clean and environmentalist product development provides competitive advantage; developing and executing

Table 7. Weights of alternative strategy

S1	0.204
S2	0.364
S3	0.238
S4	0.198
	1.003

activities for the design of green-focused products; taking package-reducing and recycling measures in design; selection of sustainable, recyclable and renewable material and production input are determined as S4(the fourth strategy).

Table 7 shows the weight of the strategies developed on the basis of SWOT factors:

- The weight value was calculated as 20.4% for (S1) Developing systems to optimize material transportation within the factory, working to reduce fuel consumption in product/service deliveries, and the green workforce that can apply technology.
- The weight value was calculated as 36.4% for (S2) specifying and determining alternatives rather than toxic chemicals; taking measures to reduce greenhouse gas emissions, reducing water consumption and saving water, and setting a target of reducing energy use.
- The weight value was calculated as 23% for (S3) Assuming that consumers will prefer green products and pay a high price for them if necessary; using eco-labeling and achieving competitive advantage; positive effect of environment-oriented measures on competitiveness; the application of the ISO 14001 environmental management standard, and the management's support and reward strategy for practices that lead to green change.
- The weight value was calculated as 19.8%. for (S4) reducing costs through resource efficiency; reducing costs by decreasing the waste amount; ensuring that green, clean, environmentalist product development provides competitive advantage; developing and executing activities for the design of green-focused products; taking package-reducing and recycling measures in design; and the selection of a sustainable, recyclable and renewable material as production input.

The inconsistency rate was calculated as 0.032. This ratio indicates that the matrix is consistent. As a result of the evaluations, S2 was determined as the most appropriate strategy.

Based on this result, the most highly weighed value due to the resolution of the priority strategy model (S2) which needs to be followed by the Textile Industry regarding its environmental sustainability is labelled (S2) as "Specifying and determining alternatives rather than toxic chemicals, taking measures to reduce greenhouse gas emissions, reducing water consumption and saving water, and setting a target to reduce energy use."

CONCLUSION

A strategic Model to provide the sustainability of garment and textile sector can be connected with these four topics. These are Government Impact, Consumer Affect (end-consumer behavior), supply chain construction- chain of events depends on the production and alterations stages, Design new products and choice the raw material of products by assorted innovation, technology. The impacts of the textile, garment sector are classified due to the sustainability categories. These are PESTLECH (political, economic, social, technological, legal, environmental, cultural and historical factors) as a holistic approach in a strategic model of sustainability.

REFERENCES

BCST. (2005). *National Research Council Board on Chemical Sciences and Technology, Sustainability in the Chemical Industry: GrandChallenges and Research Needs*. The National Academies Press. Available at http://www.nap.edu/catalog.php?record_id=11437

Bernroider, E. (2002). Factors in SWOT Analysis Applied to Micro, Small- to medium, and Large Software Enterprises: An Austrian Study. *European Management Journal*, *20*(5), 562–573. doi:10.1016/S0263-2373(02)00095-6

Bouten, L., Everaert, P., Liedekerke, L. V., & De Moor, L. (2011). Corporate Social Responsibility Reporting: A Comprehensive Picture. *Accounting Forum*, *35*(3), 187–204. doi:10.1016/j.accfor.2011.06.007

Cebeci, U. (2009). Fuzzy AHP-based decision support system for selecting ERP systems in textile industry by using balanced scorecard. *Expert Systems with Applications*, *36*(5), 8900–8909. doi:10.1016/j.eswa.2008.11.046

Chen, J.-K., & Chen, I.-S. (2010). Using a novel conjunctive MCDM approach based on DEMATEL, fuzzy ANP and TOPSIS as an innovation support system for Taiwanese higher education. *Expert Systems with Applications*, *37*(3), 1981–1990. doi:10.1016/j.eswa.2009.06.079

Defra. (2008). *A framework for pro-environmental behaviors*. London: Department for Environment, Food and Rural Affairs Report. Available at https://www.gov.uk/government/uploads/system/uploads/attachment_data/file/69277/pb13574-behaviours-report-080110.pdf

Dinçer, Ö. (2004). *Stratejik Yönetim ve İşletme Politikası. 7*. İstanbul: Baskı, Beta Yayınları.

Dyson, R. G. (2004). Strategic development and SWOT analysis at the University of Warwick. *European Journal of Operational Research*, *152*(3), 631–640. doi:10.1016/S0377-2217(03)00062-6

Expert Choıce. (2000). *Expert Choice, Analytical Hierarchy Process (AHP) Software. Version 9.5*. Expert Choice.

Garuti, C., & Spencer, I. (2007). Parallels between the analytic hierarchy and network processes (AHP/ANP) and fractal geometry. *Mathematical and Computer Modelling*, *46*(7-8), 926–934. doi:10.1016/j.mcm.2007.03.029

Harrison, J. P. (2010). *Strategic planning and Swot Analysis, Essentials of Strategic Planning in Healthcare*. Health Administration Press. Available at http://www.ache.org/pdf/secure/gifts/harrison_chapter5.pdf

Kahraman, C., Demirel, N. Ç., & Demirel, T. (2007). Prioritization of e-Government strategies using a SWOT-AHP analysis: The case of Turkey. *European Journal of Information Systems*, *16*(3), 284–298. doi:10.1057/palgrave.ejis.3000679

Kajanus, M., Kangas, J., & Kurttıla, M. (2004). The use of value focused thinking and the A SWOT hybrid method in tourism management. *Tourism Management*, *25*(4), 499–506. doi:10.1016/S0261-5177(03)00120-1

Kurtilla, M., Pesonen, M., Kangas, J., & Kajanus, M. (2000). Utilizing the analytic hierarchy process (AHP) in SWOT analysis - a hybrid method and its application to a forest certification case. *Forest Policy and Economics*, *1*(1), 41–52. doi:10.1016/S1389-9341(99)00004-0

Lee, A. H. I., Chen, W.-C., & Chang, C.-J. (2008). A fuzzy AHP and BSC approach for evaluating performance of IT department in the manufacturing industry in Taiwan. *Expert Systems with Applications*, *34*(1), 96–107. doi:10.1016/j.eswa.2006.08.022

Lee, S. F., & Lo, K. K. (2003). e-Enterprise and management course development using strategy formulations framework for vocational education. *Journal of Materials Processing Technology*, *139*(1-3), 604–612. doi:10.1016/S0924-0136(03)00501-6

Leskınen, L. A., Leskınen, P., Kurttıla, M., Kangas, J., & Kajanus, M. (2006). Adapting modern strategic decision support tools in the participatory strategy process-a case study of a forest research station. *Forest Policy and Economics*, *8*(3), 267–278. doi:10.1016/j.forpol.2004.06.007

Masozera, M. K., Alavalapatı, J. R. R., Jacobson, S. K., & Shresta, R. K. (2004). *Assessing the suitability of community-based management for the Nyungwe Forest Reserve*. Forest Policy and Economics.

Maxwell, D. (2008). Sustainable Clothing Action Plan, Department for Environment. *Lead Sustainable Clothing Roadmap, Food and Rural Affairs*, 2-11. Available at http://glasaaward.org/wp-content/uploads/2014/01/Clothing-Actionplan.pdf

OECD. (2001). *OECD Environmental Strategy for the First Decade of the 21st Century*. Retrieved from http://www.oecd.org/environment/environmentalindicatorsmodellingandoutlooks/1863539.pdf

Pan, N.-F. (2008). Fuzzy AHP approach for selecting the suitable bridge construction method. *Automation in Construction*, *17*(8), 958–965. doi:10.1016/j.autcon.2008.03.005

Promentilla, M. A. B., Furuichi, T., Ishii, K., & Tanikawa, N. (2008). A fuzzy analytic network process for multi-criteria evaluation of contaminated site remedial countermeasures. *Journal of Environmental Management*, *88*(3), 479–495. doi:10.1016/j.jenvman.2007.03.013 PMID:17467879

Saaty, T. L. (1980). *The Analytic Hierarchy Process*. McGraw-Hill International Book Company.

Saaty, T. L. (1991). Some Mathematical Concepts of the Analytic Hierarchy Process. *Behaviormetrika*, *29*(29), 1–9. doi:10.2333/bhmk.18.29_1

Saaty, T. L. (2007). Time dependent decision-making; dynamic priorities in the AHP/ANP: Generalizing from points to functions and from real to complex variables. *Mathematical and Computer Modelling*, *46*(7-8), 860–891. doi:10.1016/j.mcm.2007.03.028

Saaty, T. L., & Sodenkamp, M. (2008). Making decisions in hierarchic and network systems. *International Journal Applied Decision Sciences, 1*(1), 24-56. available at http://www.bitstoenergy.ch/wpcontent/uploads/2016/01/2008_IJADS_Saaty_Sodenkamp.pdf

Seker, S., & Özgürler, M. (2012). Analysis of the Turkish consumer electronics firm using SWOT-AHP method. *Procedia: Social and Behavioral Sciences*, *58*, 1544–1554. doi:10.1016/j.sbspro.2012.09.1141

Shaver, J. A. (1999). Applying ISO 14001 to environmental management. *Textile Chemist and Colorist*, 27–30.

Shrestha, R. K., Alavalapati, J. R. R., & Kalmbacher, R. S. (2004). Exploring the potential for silvopasture adoption in south-central Florida: An application of SWOT–AHP method. *Agricultural Systems*, *81*(3), 185–199. doi:10.1016/j.agsy.2003.09.004

Stewart, R., Moarmed, S., & Daet, R. (2002). Strategic implementation of IT/IS projects in construction: A case study. *Automation in Construction*, *11*(6), 681–694. doi:10.1016/S0926-5805(02)00009-2

Turkey's Sustainable Development Report: Claiming the Future Best Practice Examples on Sustainable Development. (n.d.). Available at http://www.undp.org/content/dam/turkey/docs/Publications/EnvSust/2.Best%20Practice%20Examples.pdf

Ülgen, H., & Mirze, S. K. (2004). *İşletmelerde Stratejik Yönetim*. İstanbul: Literatür Yayınları.

Vahidnia, M. H., Alesheikh, A. A., & Alimohammadi, A. (2009). Hospital site selection using fuzzy AHP and its derivatives. *Journal of Environmental Management*, *90*(10), 3048–3056. doi:10.1016/j.jenvman.2009.04.010 PMID:19477577

Yang, C.-L., Chuang, S.-P., & Huang, R.-H. (2009). Manufacturing evaluation system based on AHP/ANP approach for wafer fabricating industry. *Expert Systems with Applications*, *36*(8), 11369–11377. doi:10.1016/j.eswa.2009.03.023

Yüksel, İ., & Dağdeviren, M. (2007). Using the analytic network process (ANP) in a SWOT analysis – A case study for a textile firm. *Information Sciences*, *177*(16), 3364–3382. doi:10.1016/j.ins.2007.01.001

Yüksel, İ., & Dağdeviren, M. (2010). Using the fuzzy analytic network process (ANP) for Balanced Scorecard (BSC): A case study for a manufacturing firm. *Expert Systems with Applications*, *37*(2), 1270–1278. doi:10.1016/j.eswa.2009.06.002

APPENDIX

Annex 1: Binary Comparison Matrices of SWOT Group Sub Factors

Table: Binary Comparison Matrices of Sub-Factors in the *Opportunities Group* and Local and General Weights

Table: Binary Comparison Matrices of Sub-Factors in the *Threats Group* and Local and General Weights

Table: Binary Comparison Matrices of Sub-Factors in the *Opportunities Group* and Local and General Weights

Table: Binary Comparison Matrices of Sub-Factors in the *Weaknesses Group* and Local and General Weights

Annex 2: Binary Comparison Matrices of Alternative Strategies Based on Sub-Factors of Opportunities (O) Group

Table: Binary Comparison Matrix and Local and General Weights for Strategies according to *O1 Factor*

Table: Binary Comparison Matrix and Local and General Weights for Strategies according to *O2 Factor*

Table: Binary Comparison Matrix and Local and General Weights for Strategies according to *O3 Factor*

Table: Binary Comparison Matrix and Local and General Weights for Strategies according to *O4 Factor*

Annex 3: Binary Comparison Matrices of Alternative Strategies Based on Sub-Factors of Threats (T) Group

Table: Binary Comparison Matrix and Local and General Weights for Strategies according to *T1 Factor*

Table: Binary Comparison Matrix and Local and General Weights for Strategies according to *T2 Factor*

Table: Binary Comparison Matrix and Local and General Weights for Strategies according to *T3 Factor*

Table: Binary Comparison Matrix and Local and General Weights for Strategies according to *T4 Factor*

Table 8.

STRENGTHS				
	S1	S2	S3	S4
S1	1	0,5	2,4	0,3333333
S2	2	1	0,5	0,16
S3	0,47	2	1	0,16
S4	5	6	6	1
Total	8,47	9,5	9,9	1,6533333

STRENGTHS					
	S1	S2	S3	S4	WEIGHT
S1	0,118	0,053	0,242	0,202	0,154
S2	0,236	0,105	0,051	0,097	0,122
S3	0,055	0,211	0,101	0,097	0,116
S4	0,590	0,632	0,606	0,605	0,608
NORMALIZATION					

WEAKNESSES				
	W1	W2	W3	W4
W1	1	0,5	0,142	0,5
W2	2	1	0,3333333	0,142
W3	7	3	1	3
W4	2	7	0,3333333	1
Total	12	11,5	1,8086667	4,642

WEAKNESSES					
	W1	W2	W3	W4	WEIGHT
W1	0,083	0,043	0,079	0,108	0,078
W2	0,167	0,087	0,184	0,031	0,117
W3	0,583	0,261	0,553	0,646	0,511
W4	0,167	0,609	0,184	0,215	0,294
NORMALIZATION					

OPPORTUNITIES				
	O1	O2	O3	O4
O1	1	3	0,2	0,2
O2	0,3333333	1	5	1
O3	5	0,2	1	0,5
O4	5	1	2	1
Total	11,333333	5,2	8,2	2,7

OPPORTUNITIES					
	O1	O2	O3	O4	WEIGHT
O1	0,088	0,577	0,024	0,074	0,191
O2	0,029	0,192	0,610	0,370	0,300
O3	0,441	0,038	0,122	0,185	0,197
O4	0,441	0,192	0,244	0,370	0,312
NORMALIZATION					

THREATS				
	T1	T2	T3	T4
T1	1	2	2	2
T2	0,47	1	3	5
T3	0,5	0,3333333	1	1
T4	0,47	0,2	1	1
Total	2,44	3,5333333	7	9

THREATS					
	T1	T2	T3	T4	WEIGHT
T1	0,410	0,566	0,286	0,222	0,371
T2	0,193	0,283	0,429	0,556	0,365
T3	0,205	0,094	0,143	0,111	0,138
T4	0,193	0,057	0,143	0,111	0,126
NORMALIZATION					

Annex 4: Binary Comparison Matrices of Alternative Strategies Based on Sub-Factors of Strengths (S) Group

Table: Binary Accrual Matrix and Local and General Weights for Strategies according to *S1 Factor*

Table: Binary Accrual Matrix and Local and General Weights for Strategies according to *S2 Factor*

Table: Binary Accrual Matrix and Local and General Weights for Strategies according to *S3 Factor*

Table: Binary Accrual Matrix and Local and General Weights for Strategies according to *S4 Factor*

Table 9.

OPPORTUNITIES				
O1	S1	S2	S3	S4
S1	1	5	0,2	0,5
S2	0,2	1	0,3333333	4
S3	5	3	1	5
S4	2	0,25	0,2	1
Total	8,2	9,25	1,7333333	10,5

OPPORTUNITIES					
O1	S1	S2	S3	S4	WEIGHT
S1	0,1220	0,5405	0,1154	0,0476	0,2064
S2	0,0244	0,1081	0,1923	0,3810	0,1764
S3	0,6098	0,3243	0,5769	0,4762	0,4968
S4	0,2439	0,0270	0,1154	0,0952	0,1204
NORMALIZATION					

OPPORTUNITIES				
O2	S1	S2	S3	S4
S1	1	3	0,2	0,33
S2	0,333	1	0,33	0,2
S3	5	3	1	0,17
S4	3	5	6	1
Total	9,333	12	7,53	1,7

OPPORTUNITIES					
O2	S1	S2	S3	S4	WEIGHT
S1	0,1071	0,2500	0,0266	0,1941	0,1445
S2	0,0357	0,0833	0,0438	0,1176	0,0701
S3	0,5357	0,2500	0,1328	0,1000	0,2546
S4	0,3214	0,4167	0,7968	0,5882	0,5308
NORMALIZATION					

OPPORTUNITIES				
O3	S1	S2	S3	S4
S1	1	5	0,2	0,5
S2	0,2	1	0,33	4
S3	5	3	1	5
S4	2	0,25	0,2	1
Total	8,2	9,25	1,73	10,5

OPPORTUNITIES					
O3	S1	S2	S3	S4	WEIGHT
S1	0,1220	0,5405	0,1156	0,0476	0,2064
S2	0,0244	0,1081	0,1908	0,3810	0,1761
S3	0,6098	0,3243	0,5780	0,4762	0,4971
S4	0,2439	0,0270	0,1156	0,0952	0,1204
NORMALIZATION					

OPPORTUNITIES				
O4	S1	S2	S3	S4
S1	1	0,2	2	2
S2	5	1	7	5
S3	0,5	0,1428571	1	2
S4	0,5	0,2	0,5	1
Total	7	1,5428571	10,5	10

OPPORTUNITIES					
O4	S1	S2	S3	S4	WEIGHT
S1	0,1429	0,1296	0,1905	0,2000	0,1657
S2	0,7143	0,6481	0,6667	0,5000	0,6323
S3	0,0714	0,0926	0,0952	0,2000	0,1148
S4	0,0714	0,1296	0,0476	0,1000	0,0872
NORMALIZATION					

Annex 5: Binary Comparison Matrices of Alternative Strategies Based on Sub-Factors of Weaknesses (W) Group

Table: Binary Accrual Matrix and Local and General Weights for Strategies according to *W1 Factor*

Table: Binary Accrual Matrix and Local and General Weights for Strategies according to *W2 Factor*

Table: Binary Accrual Matrix and Local and General Weights for Strategies according to *W3 Factor*

Table: Binary Accrual Matrix and Local and General Weights for Strategies according to *W4 Factor*

Table 10.

THREATS				
T1	S1	S2	S3	S4
S1	1	0,5	0,125	0,5
S2	2	1	0,3333333	0,25
S3	8	3	1	7
S4	2	4	0,1428571	1
Total	18	1,4285714	8,45	12,2

THREATS					
T1	S1	S2	S3	S4	WEIGHT
S1	0,0556	0,1000	0,0237	0,0164	0,0489
S2	0,3889	0,7000	0,8284	0,5738	0,6228
S3	0,2778	0,1000	0,1183	0,3279	0,2060
S4	0,2778	0,1000	0,0296	0,0820	0,1223
NORMALIZATION					

THREATS				
T2	S1	S2	S3	S4
S1	1,000	7,000	9,000	2,000
S2	0,143	1,000	0,167	3,000
S3	0,111	6,000	1,000	5,000
S4	0,500	0,333	0,200	1,000
Total	1,75397	14,33333	10,36667	11,00000

THREATS					
T2	S1	S2	S3	S4	WEIGHT
S1	0,5701	0,4884	0,8682	0,1818	0,5271
S2	0,0814	0,0698	0,0161	0,2727	0,1100
S3	0,0633	0,4186	0,0965	0,4545	0,2582
S4	0,2851	0,0233	0,0193	0,0909	0,1046
NORMALIZATION					

THREATS				
T3	S1	S2	S3	S4
S1	1	0,5	0,2	0,2
S2	2	1	0,3333333	1
S3	5	3	1	2
S4	5	1	0,5	1
Total	13	5,5	2,0333333	4,2

THREATS					
T3	S1	S2	S3	S4	WEIGHT
S1	0,0769	0,0909	0,0984	0,0476	0,0785
S2	0,1538	0,1818	0,1639	0,2381	0,1844
S3	0,3846	0,5455	0,4918	0,4762	0,4745
S4	0,3846	0,1818	0,2459	0,2381	0,2626
NORMALIZATION					

THREATS				
T4	S1	S2	S3	S4
S1	1,00	0,14	0,20	0,20
S2	7,00	1,00	7,00	7,00
S3	5,00	0,14	1,00	4,00
S4	5,00	0,14	0,25	1,00
Total	13,000	8,500	1,601	8,750

THREATS					
T4	S1	S2	S3	S4	WEIGHT
S1	0,0769	0,0588	0,0781	0,0571	0,0677
S2	0,1538	0,1176	0,2082	0,0286	0,1271
S3	0,6154	0,3529	0,6245	0,8000	0,5982
S4	0,1538	0,4706	0,0892	0,1143	0,2070
NORMALIZATION					

Table 11.

STRENGTHS(ST)				
ST1	S1	S2	S3	S4
S1	1,0	1,0	0,3	0,3
S2	1,0	1,0	0,3	0,2
S3	4,0	3,0	1,0	1,0
S4	3,0	5,0	1,0	1,0
Total	9,0	10,0	2,6	2,5

STRENGTHS(ST)					
ST1	S1	S2	S3	S4	WEIGHT
S1	0,1111	0,1000	0,0968	0,1315	0,1098
S2	0,1111	0,1000	0,1289	0,0790	0,1047
S3	0,4444	0,3000	0,3871	0,3948	0,3816
S4	0,3333	0,5000	0,3871	0,3948	0,4038
NORMALIZATION					

STRENGTHS(ST)				
ST2	1,0	0,3	0,1	0,1
S1	3,0	1,0	0,3	0,1
S2	7,0	3,0	1,0	1,0
S3	9,0	7,0	1,0	1,0
S4	0,3	0,3	0,5	1,0
Total	19,25	11,33	2,83	3,14

STRENGTHS(ST)					
ST2	S1	S2	S3	S4	WEIGHT
S1	0,1558	0,0882	0,1176	0,0455	0,1018
S2	0,3636	0,2647	0,3529	0,3182	0,3249
S3	0,4675	0,6176	0,3529	0,3182	0,4391
S4	0,0130	0,0294	0,1765	0,3182	0,1343
NORMALIZATION					

STRENGTHS(ST)				
ST3	S1	S2	S3	S4
S1	1,0	0,5	5,0	4,0
S2	2,0	1,0	1,0	3,0
S3	0,2	1,0	1,0	2,0
S4	0,3	0,3	0,5	1,0
Total	3,5	2,8	7,5	10,0

STRENGTHS(ST)					
ST3	S1	S2	S3	S4	WEIGHT
S1	0,2899	0,1767	0,6667	0,4000	0,3833
S2	0,5797	0,3534	0,1333	0,3000	0,3416
S3	0,0580	0,3534	0,1333	0,2000	0,1862
S4	0,0725	0,1166	0,0667	0,1000	0,0889
NORMALIZATION					

STRENGTHS(ST)				
ST4	S1	S2	S3	S4
S1	1	1	0,333	0,5
S2	1	1	0,2	0,14
S3	3	5	1	1
S4	2	7	1	1
Total	7	14	2,533	2,64

STRENGTHS(ST)					
ST4	S1	S2	S3	S4	WEIGHT
S1	0,1429	0,0714	0,1315	0,1894	0,1338
S2	0,1429	0,0714	0,0790	0,0530	0,0866
S3	0,4286	0,3571	0,3948	0,3788	0,3898
S4	0,2857	0,5000	0,3948	0,3788	0,3898
NORMALIZATION					

Table 12.

WEAKNESSES				
W1	S1	S2	S3	S4
S1	1	3	5	1
S2	0,333	1	5	7
S3	0,2	0,2	1	1
S4	1	0,14	1	1
Total	2,533	4,34	12	10

WEAKNESSES					
W1	S1	S2	S3	S4	WEIGHT
S1	0,3948	0,6912	0,4167	0,1000	0,4007
S2	0,1315	0,2304	0,4167	0,7000	0,3696
S3	0,0790	0,0461	0,0833	0,1000	0,0771
S4	0,3948	0,0323	0,0833	0,1000	0,1526
NORMALIZATION					

WEAKNESSES				
W2	S1	S2	S3	S4
S1	1	5	0,2	0,5
S2	0,2	1	0,3333333	4
S3	5	3	1	5
S4	2	0,25	0,2	1
Total	8,2	9,25	1,7333333	10,5

WEAKNESSES					
W2	S1	S2	S3	S4	WEIGHT
S1	0,1220	0,5405	0,1154	0,0476	0,2064
S2	0,0244	0,1081	0,1923	0,3810	0,1764
S3	0,6098	0,3243	0,5769	0,4762	0,4968
S4	0,2439	0,0270	0,1154	0,0952	0,1204
NORMALIZATION					

WEAKNESSES				
W3	S1	S2	S3	S4
S1	1	0,3333333	0,3333333	0,3333333
S2	3	1	5	5
S3	3	0,2	1	2
S4	3	0,2	0,5	1
Total	10	1,7333333	6,8333333	8,3333333

WEAKNESSES					
W3	S1	S2	S3	S4	WEIGHT
S1	0,1000	0,1923	0,0488	0,0400	0,0953
S2	0,3000	0,5769	0,7317	0,6000	0,5522
S3	0,3000	0,1154	0,1463	0,2400	0,2004
S4	0,3000	0,1154	0,0732	0,1200	0,1521
NORMALIZATION					

WEAKNESSES				
W4	S1	S2	S3	S4
S1	1	2	5	3
S2	0,5	1	7	5
S3	0,2	0,1428571	1	1
S4	0,3333333	0,2	1	1
Total	2,0333333	3,3428571	14	10

WEAKNESSES					
W4	S1	S2	S3	S4	WEIGHT
S1	0,4918	0,5983	0,3571	0,3000	0,4368
S2	0,2459	0,2991	0,5000	0,5000	0,3863
S3	0,0984	0,0427	0,0714	0,1000	0,0781
S4	0,1639	0,0598	0,0714	0,1000	0,0988
NORMALIZATION					

Chapter 6
Strategic Implementation of Green Public Procurement in the City of Buenos Aires

Leslie Elizabeth Harper
Inter-American Development Bank, USA

Daniel Sanchez
Inter-American Development Bank, USA

ABSTRACT

As public procurement systems around the world continue their modernization process, many countries are including a new objective in their reform efforts: the implementation of public policies in the procurement process. This has been a growing trend in Latin America, where countries have shown leadership in incorporating public policies into their modernization plans, being a perfect example is the introduction of green public procurement standards. There is particular interest in incorporating sustainability in public procurement at the sub-national level where governments are directly responsible for the provision of citizen services. In the case of Argentina, the Government of the City of Buenos Aires is at the forefront of public procurement modernization, leading the country in green public procurement strategies. This chapter explores the principles and global trends that have influenced the City of Buenos Aires in implementing green public procurement and provides insight on the economic rationale and potential benefits for one of the largest metropolitan regions in LAC.

DOI: 10.4018/978-1-5225-7083-7.ch006

INTRODUCTION

Promoting environmental sustainability is a mainstay of any National Development Program in Latin America and the Caribbean (LAC). Among many approaches to implement this priority, incorporating sustainability standards to public procurement (Green Public Procurement-GPP) has become relevant because of its potential to support the achievement of this policy objective.

Government expenditure in LAC countries still constitutes a significant part of GDP, thus its potential to create incentives to promote sustainable economic sectors and corresponding practices. Currently, the adoption and implementation of GPP standards in public procurement legislation is an inherent part of any reform effort in the LAC region.

This has not only been happening at the national level. Sub-national governments such as states, provinces and municipalities throughout the region have also been increasingly active in incorporating GPP practices into their public procurement as the benefits are more immediate and direct at this level. Municipalities and local governments have a great potential for positive environmental impact as they are usually in charge of procurement of services such as public transportation, electricity and garbage disposal, which have significant scope for implementing GPP criteria. A key example of this is the City of Buenos Aires, which stands out as one of the pioneer cities to implement GPP in Argentina and the LAC region.

PRINCIPLES OF GPP THAT INFLUENCED THE CITY OF BUENOS AIRES

Governments, international organizations and think tanks have developed a variety of different definitions of GPP, in many cases included within the overall broader framework of sustainable procurement. However, there are common concepts and definitions across all of them, which include Value for Money (VfM), life cycle of goods and services, and environmental sustainability over time in the provision of goods and services.

DEFINITIONS OF GPP

Some definitions that have served as reference to the City of Buenos Aires are:

Figure 1. GPP policy objectives
Source: Author's own analysis with UNEP and EU Commission data.

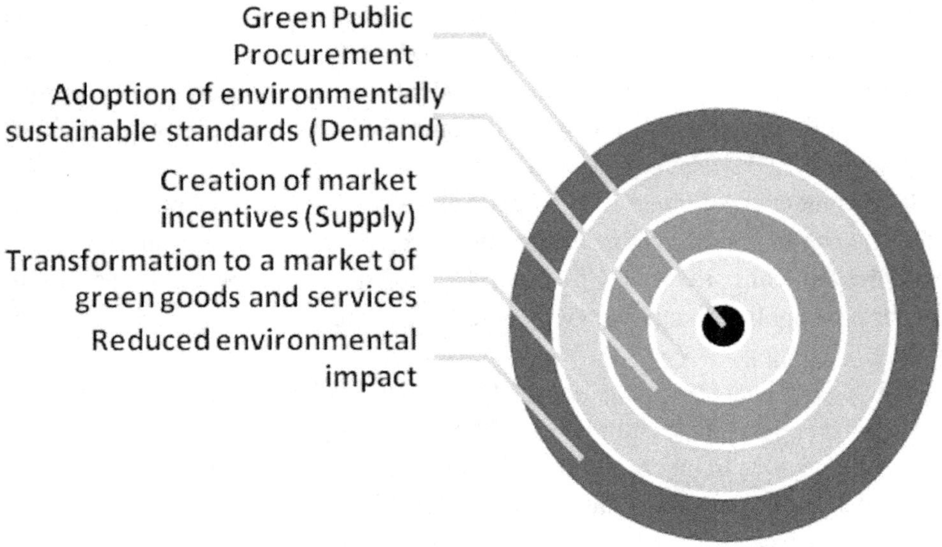

- **European Union:** "Green Procurement is a process whereby public authorities seek to procure goods, services and works with a reduced environmental impact throughout their life cycle when compared to goods, services and works with the same primary function that would otherwise be procured." (EU Commission, 2008)
- **United Nations:** "Sustainable procurement is about combining social and environmental factors with financial considerations when making purchasing decisions. It involves looking beyond the traditional economic parameters and making decisions based on life-cycle costs, associated environmental and social risks and benefits as well as broader environmental implications." (UNEP, 2008)
- **United States Government:** "Sustainability is best thought of as a change process, rather than an end state. Sustainability means to create and maintain conditions, under which humans and nature can exist in productive harmony that permits fulfilling the social, economic, and other requirements of present and future generations." (GSA, 2007)
- **Canadian Government:** "A lesser or reduced impact on the environment over the life cycle of the good or service, when compared with competing goods or services serving the same purpose." (PSPC, 2006)

Sectors and Trends for Implementing GPP

Although there are numerous ways to interpret and implement these principles, green procurement strategies for the most part aim to "reduce the consumption of resources, utilities and energy, avoid waste and pollutant emissions, increase quality, protect biodiversity, reduce internal and external environmental costs (e.g. costs for disposal or transport), increase the transparency and plausibility of costs, foster innovation, fair working conditions and income, create "green jobs" and win suppliers as strategic partners."(OECD, 2016).

GPP articulates demand and supply in order to use a government's purchasing power to create incentives for the production of goods without negative effects for the environment.

Benefits of Green Procurement

Given that public procurement accounts for approximately 15-20% of GDP, and up to 30% in many developing countries, the potential impact that having green policies can have is enormous, affecting not just the immediate purchase of a good or service, but causing a rippling effect throughout the economy and how society views, buys and sells green products over time. It can affect the environment itself, lower costs, improve the image of the government and spur innovation in the private sector. (Casier et al., 2015)

Table 1. Areas of frequent public expenditure where GPP can be implemented

Goods	Services	Works
Air conditioning	Courier and postal	Airports
Apparel	Data centers	Building
Fuel	Electricity	Ports
Furniture	Food and beverage	Power plants
Health supplies	IT Consultancy	Prisons
Information technology	Landscaping	Railroads
Lighting	Management Consultancy	Roads
Office supplies	Mobility	Schools
Paper	Software	Sewage
Vehicles	Waste management	Water treatment

Source: OEDC (2016).

Figure 2. Share of Public Procurement as Percentage of GDP
Source: SELA (2015).

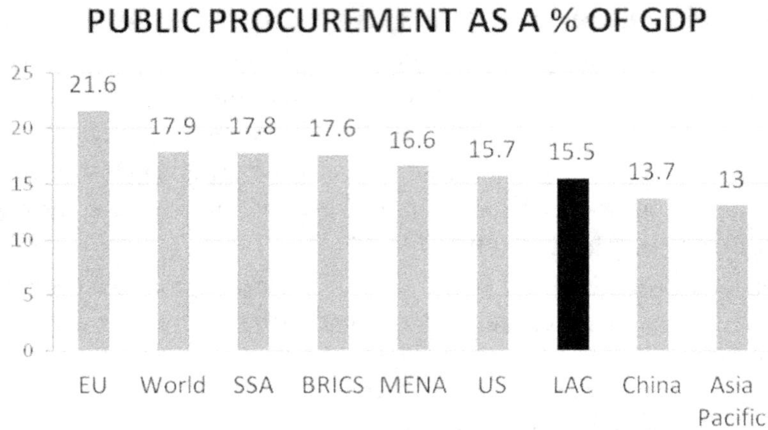

Additionally, there is still considerable scope for improvement in this area in LAC, for example only 40% of the legislation on public procurement includes some type of green policies, and only 19% of the regulations apply the concept of sustainability in public procurement (SELA, 2015).

Some of the benefits of GPP relevant for LAC and therefore the City of Buenos Aires are:

Environmental Benefits

The environmental impacts of green procurement are myriad and can help mitigate problems such as deforestation by purchasing wood from sustainably managed forests. Greenhouse gas emissions and excess water use can be curbed by procuring energy efficient products (ie. green light bulbs). Finally, waste can be reduced by taking into account disposal implications of goods purchased (Commission of the European Communities, 2016). For example, in Europe, the EU estimates that "CO2 emissions would be cut by 15 million tons per year if the whole EU adopted the same environmental criteria for lighting and office equipment as the City of Turku, Finland - reducing electricity consumption by 50%". (OECD, 2015)

Economic Benefits

Contrary to popular belief, buying green is not necessarily more expensive. In fact, it can save governments money over time as green products use less energy, do not lead to as much waste and last longer. Green procurement does this as at its very core it leads to a change in approach to entire cost of ownership and not just the purchase price but also the true cost over time, which includes operations, repair and replacement costs, management and disposal, among others (Sieb, 2011).

There are multiple ways that green public procurement provides economic benefits. The first is through savings over the medium and long term when life-cycle costs are taken into account. For example, a more energy efficient building may cost more initially but will cost less to heat and cool than a non-energy efficient building. Products that are easily recycled, have less packaging or last longer will result in reduced waste (IISD, 2013). Another example that is often used is that of light bulbs – if you consider that Compact Fluorescent bulbs last on average 10 years longer than regular light bulbs and are more energy efficient you have savings from reduced use of energy, lower maintenance costs as workers are not needed as often to change them and lower costs for their replacements. (Strandberg & Robinson, 2008)

Another way is through decreased cost of green products and services. Over time, as the market for green products and services deepens, those initial higher costs should become more comparable as more companies enter the market and competition increases. This can support the development of incentives for the private sector to innovate in green products and services. This is especially important for small and medium sized enterprises as it can provide a niche for their products, which can benefit from the market power of the public sector (Commission of the European Communities, 2016). This impact on the market place is another benefit that the public sector can provide as one of the largest purchasers in a country. (EU Commission, 2016)

Finally, savings can also be found in a reduction in legal and insurance costs, which can occur after a hazardous materials incident, health and safety claims and violations or other problems that necessitate legal counsel or penalties. (Strandberg & Robinson, 2008)

Social Benefits

Green procurement has benefits beyond just that of improving the environment. By implementing policies that aim to ensure better quality of goods and services, better public health and occupational health benefits can be achieved. (OECD, 2015). For example, by promoting energy efficient buildings with natural ventilation or by procuring services such as cleaners, solvents and paints with decreased toxic chemicals. Air quality in public procurement can be improved through changing building materials such as carpet, furniture or wall coverings. A focus on renewable energy could decrease cancers and respiratory problems from fuel extraction and air pollution. (WHO, 2014) For example, a study in Washington State demonstrated that each worker's compensation claim related to cleaning products meant the worker was off the job for an average of 18 hours with a cost per worker of $725 (including lost work time) and that six out of every 100 workers are injured as a consequence of harmful cleaners. (Strandberg & Robinson, 2008)

Green public procurement can also encourage higher quality standards for products and services which can result in healthier working environments with a spillover effect to private consumers as those types of products become in higher demand. (OECD, 2015)

Another social benefit of green procurement is with regard to job creation. In Korea for example, the Ministry of Environment introduced the Act on Encouragement of Purchase of Green Products in 2005. Since then, it is estimated that 12,143 new jobs have been created. (OECD, 2015).

Market Benefits

As the public procurement sector is so large, policies that encourage green procurement can have an enormous impact on the private sector. By creating a large public market for green products, the private sector is encouraged to produce and sell them in the open market. By creating an economy of scale, this makes it easier for them to produce and sell in the private sector as well. In addition, by encouraging innovation, suppliers develop products which they wouldn't have otherwise, which can potentially be scaled up to other buyers.

Political Benefits

In many countries, there is increased concern for the environment and the impact that the public sector has on a given country. By going green, governments not only raise environmental awareness among the public, but also show that they are doing their part through initiatives that will create a healthier environment for its citizens. In addition, the potential savings from green procurement, particularly those gained from taking in account the entire life cycle cost of a product, can be measured leaving resources available for other key public-sector initiatives which can be useful to policy makers.

ECONOMIC AND COST PRINCIPLES FOR GPP IN LAC

Life-Cycle Costing

Life Cycle costing can be defined as an "economic assessment considering all agreed projected significant and relevant cost flows over a period of analysis expressed in monetary value. The projected costs are those needed to achieve defined levels of performance, including reliability, safety and availability." (Perera, 2009) The idea behind life cycle costing is that purchasers go beyond just the cost of the item at purchase but rather all costs of the product including delivery, installation, commissioning and insurance; operating costs, including utility costs and maintenance costs; end-of-life costs such as removal, recycling or refurbishment and disposal. This is a critical component when taking into account VfM – that environmental considerations are not just considered when the good or service is being purchased but over its life cycle which include the medium and long-term. (Casier et al., 2015)

Figure 3. Typical Life-Cycle Costs to consider
Source: Authors' own analysis with IISD information.

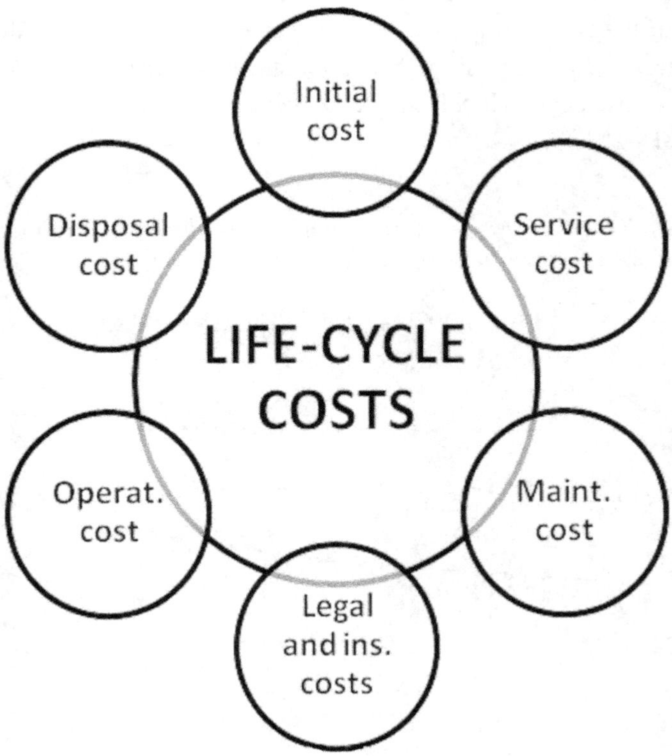

Adoption of Life-Cycle Costing in LAC

In LAC, there has been a growing understanding about the importance of considering the entire life cycle cost of a product and not the acquisition cost. This can be seen in the number of laws and policies in the region with dedicated provisions related to life cycle costs. In Uruguay for example, Energy Efficiency Law No. 18.597 mandates that procurers take into account life cycle costs including purchase price, operating cost, management of operating waste and end of life disposal. (Casier et al., 2015)

Another way that countries in the region are incorporating life cycle considerations in the procurement cycle is through eco-labels. Eco-labels are a tool which help purchasers identify green products or services and in confirming compliance. The International Organization for Standardization has two types of eco-labels related to life-cycle including Type I (ISO 14024), which provides a "life-cycle seal of approval" and Type III (ISO 14025), which provides quantified environmental data on a product's life-cycle impact on the environment, enabling an environmental performance assessment (Global Eco-labeling Network, 2012). They are particularly helpful in low capacity environments as they can also assist the procuring entity with criteria to include in technical specifications. In addition, as a third party is involved, the entity itself does not have to assess that that high environmental standards are being applied, which can require a level of capacity and expertise that a country might not have (EU Commission, 2016). Countries in LAC that have particularly emphasized the use of such labels include Argentina, Chile, Mexico, Colombia and Ecuador. (IISD, 2013)

Figure 4. Life-Cycle Cost Analysis vs. Acquisition-only Price Analysis
Source: Authors' own analysis with World Bank and IISD information.

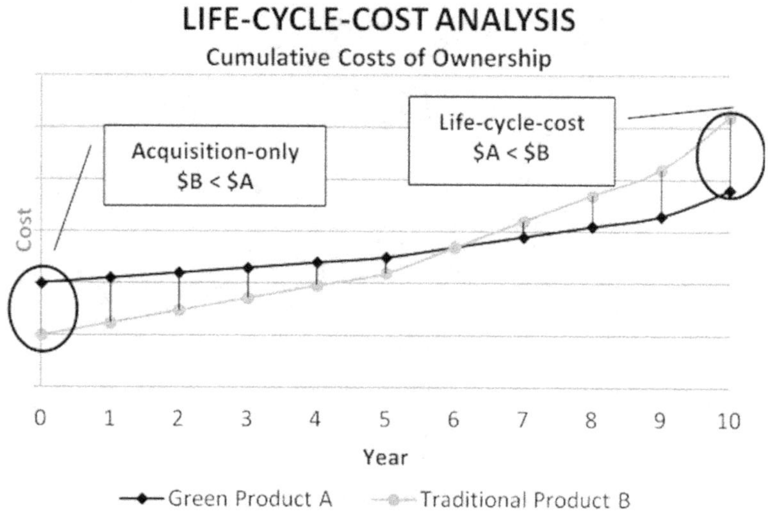

Importance of Life-Cycle Costing for GPP

Life cycle costing is a critical element of consideration for green procurement for two main reasons. The first is the impact that that product can have on the country itself. As life cycle costing takes into account the impact from start to finish of the product, it provides incentives to procure items or services that are more efficient, for example the typical example of a light bulb that is more expensive to buy in the first place but as it is more energy efficient, saves money for the government over the life span of that bulb. That means that less electricity is being used. One bulb does not make that much of a difference, but if one considers the vast number of bulbs and electricity that could potentially be saved if such bulbs were used by the entire public sector, there the impact is considerable.

The second is that as the government is often the largest purchaser in the economy, the purchasing power of the public sector can make a tremendous difference in innovations for green products and the variety and quantity that are available in the local market. When the public sector includes the total life-cycle cost of the product in the goods and services it procures, it makes it more possible in business to business procurement as well. Without the presence of the public sector in the market for these types of goods and services, it might not be economically viable for companies to produce them or sufficient incentives to invest in innovation in this area.

Challenges for Implementing Life-Cycle Costing in LAC

A main barrier for implementing life cycle costing in public procurement in LAC has been the perceived cost both in terms of the environmental assessment and the initial higher price of the product. A global survey carried out by the United Nations Environmental Program (UNEP) found that it was the number one barrier. (Yaker et al, 2013) Other challenges include lack of political support, limited capacity to implement life cycle costing in the process, particularly evaluation of green goods and services, and absence of coordination between procurement officers and those working on promoting environmental protection in the government. In many countries it is not clear who is in charge of leading the topic, which has resulted in a lack of a holistic vision and more limited and haphazard results than they would have been achieved otherwise (Grinier, 2014).

Value for Money

There are many definitions of VfM, for example the UK Treasury defines it as "the optimum combination of whole-of-life costs and quality (or fitness of purpose) of the good or service to meet a user's requirements." (World Bank Institute, 2013) This

Table 2. Products and services that are best suited for life cycle costing in procurement

Level of Applicability of Life Cycle Costing			
Frequently Purchased Items	**High**	**Medium**	**Low applicable**
GOODS			
Air conditioning	X		
Apparel	X		
Fuel	X		
Furniture		X	
Health supplies	X		
ICT equipment	X		
Lighting	X		
Office Supplies			X
Paper		X	
Vehicles	X		
SERVICES			
Catering		X	
Couriers and postal services		X	
Electricity	X		
IT Consultancy	X		
Landscaping	X		
Management Consultancy	X		
Mobility	X		
Software			X
Transport	X		
Waste management	X		
WORKS			
Landscaping		X	
New buildings	X		
Railways	X		
Refurbishment of buildings	X		
Roads	X		

Source: IISD (2013).

means that countries go beyond just focusing on price, which may not provide the best outcome, but rather to include "the degree to which a set of inherent characteristics fulfils requirements" (CIPS, 2017).

More recently however, there has been a move towards an expanded definition that includes the overall policies and goals of the government balancing "features of the desired goods/services with the cost of obtaining them as well as the political and social goals being pursued." (Williams-Elegbe, 2017) The OECD further clarifies it by "striking the best balance between the three e's - economy, efficiency and effectiveness" with another "e" that of equity sometimes being used (Jackson, 2012).

VfM definitions can be varied and in many cases complex as this analysis was primarily designed for assessing big infrastructure projects. However, for GPP purposes we can summarize the concept in the interaction of three factors:

- **Price:** The quantity or amount of payment for goods and services
- **Cost:** The expenses to be made in order to acquire a good or a service
- **Value:** Measure of the benefit that can be acquired from a good or service, or its provider

Relevance to GPP

According to the Chartered Institute of Procurement and Supply (CIPS), the determinants of VfM can vary widely depending on the objective of the public entity carrying out the procurement or the development objective pursued. These can include for example "fitness for purpose, quality, total lifetime costs, risk, environmental and sustainability issues, and a variety of other factors relating to the contribution of the proposed solutions to the organization's overall goals." (CIPS, 2012)

Therefore, VfM is relevant to green procurement as it is possible to include environmental protection as one of the factors that determine it. Governments that want to address environmental concerns as part of a government policy for the economy can limit the negative impact on the environment from actions by the public sector by implementing strategies to avoid unnecessary consumption and management of demand, choosing products that have lower adverse impacts resulting from any stage in their production, use or disposal, fostering innovation in sustainable products and services through the design and implementation of procurements. Examples of such factors can include financial savings through greater energy efficiency, lowered waste disposal costs, reduced water use and reusing materials and products (Government of the Commonwealth of Australia, 2013).

Adoption in LAC

In LAC, there has been an increased focus on adopting VfM in public expenditure. Budget constraints stemming from financial crises and lower commodity prices

Figure 5. Three dimensions of VfM
Source: Authors' own analysis with World Bank and CIPS information.

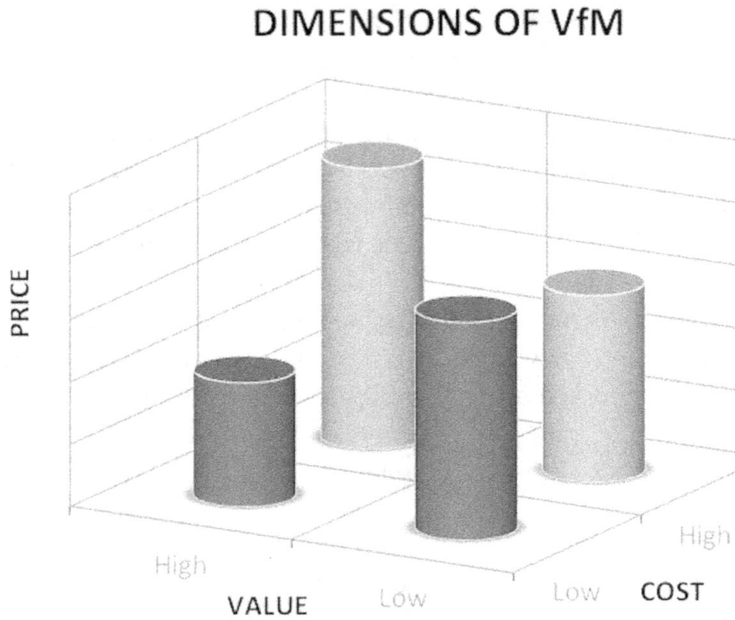

have resulted in countries prioritizing not only spending less but also making sure that the funds spent are achieving what they were supposed to do and in a way that supports the governments overall development objectives. At the same time, there has also been an increased pressure by civil society and the media on governments to do more with less and one way to do that is through a greater emphasis on VfM. Development organizations are also putting pressure on countries to demonstrate evidence that interventions have worked, and implementing VfM with corresponding evidence-based actions is a concrete way to show impact and the effectiveness of public money being spent.

GREEN PUBLIC PROCUREMENT IN THE CITY OF BUENOS AIRES

Definitions, principles and strategies mentioned in the previous section have been disseminated widely in the LAC region and have been fundamental to create momentum to introduce GPP in public procurement legislation, regulation and operations in Argentina.

Environmental sustainability and environment-friendly production have been a matter of public policy debate in Argentina for a long time, as it is one of the most industrialized countries in LAC. These concerns have been conducive to the adoption of GPP principles.

Nowhere is this more apparent than in the City of Buenos Aires where there is a commitment at the highest levels to not only make the city's procurement system efficient and transparent but also to turn it into an instrument for public policy. This includes for example the implementation of energy efficiency standards, reduction and improvement of residue management and responsible use of paper.

In the past decade, the Government of the City of Buenos Aires (GCBA) has transformed public management in Argentina, promoting innovative modalities that facilitate participation from different stakeholders in public policy design. Under this premise, the city's GPP initiative works with an inclusive approach with different institutions in the private and public sector to share information and design the tools necessary to implement GPP.

The following sections will explore the City of Buenos Aires as one of the pioneer jurisdictions in Argentina and the region to implement GPP. Insights will be provided on what are the environmental efforts in Argentina, the particularities of one of the biggest metropolitan areas in the region and finally how the City of Buenos Aires has implemented GPP practices in their public procurement processes.

The Ruling Principles of Sustainable Development in Argentina

Argentina has a solid legislative foundation for environmental protection, establishing the principles and mandate necessary to enable GPP implementation. In turn, the City of Buenos Aires has relied on these ruling principles and regulatory framework to develop its own regulations which constitute the cornerstone for the implementation of a GPP program.

Regulatory Framework

Argentina's legislative instruments for sustainable development and environmental protection leveraged by the City of Buenos Aires include:

- **The National Constitution:** Establishes ruling principles guaranteeing the right to a clean environment, apt for human development and that production activities satisfy demand without jeopardizing future generations. It also guarantees environmental preservation and restoration in case these activities produce any damage. Section 41 of the Argentina Constitution mandates public authorities to provide protection and oversee the rational use of natural resources, biodiversity preservation and environmental education. Finally, it requires regulation regarding budgeting to carry out this mandate, at the federal and sub-national levels. Section 43 enshrines environmental protection in the law, giving the judicial power the responsibility for oversight and enforcement of environmental laws and regulations, resulting in a conducive legal environment that seeks to ensure fulfillment of norms and rules within GPP processes (Constitution of the Argentine Nation, n.d.).

- **The General Environment Law Nº 25675:** Establishes minimum budgets necessary for appropriate management, preservation and protection of biodiversity. Article 4 establishes principles to be considered when applying this law; they are perfectly aligned with the definition of sustainable development: coherence, prevention, precaution, intergenerational equity, progressiveness, responsibility, subsidiarity, sustainability, solidarity and cooperation. These principles became valuable when drafting contractual clauses and recommendation guides for GPP (Argentina General Environment Law 25.675, 2002).
- **Renewable Energy Law Nº 26190 Modified by Law Nº 27191:** Establishes coordinated activities with public administration entities to promote renewable energy initiatives aimed to provide electricity services, research and development, and manufacturing of related equipment (Ministry of Energy, n.d.).
- **Electrical Emergency Decree 134/15:** Mandates that the Ministry of Energy and Mining implement necessary actions to guarantee quality and sustained provision of electricity, as well as mandates that public administration entities including decentralized bodies coordinate a program for rationing electrical consumption (Argentina, n.d.).

Procurement laws, norms and regulations for the GCBA do not only allow and promote GPP, but in many occasions, impose and mandate its implementation.

- **The City of Buenos Aires Constitution:** Affirms that private and public economic activities will be promoted within the boundaries of social wellbeing and sustainable development (Argentina, n.d.).
- **The City Procurement and Contracting Law N° 2095:** Legislates the procurement of goods and services in the City of Buenos Aires, it incorporates the main principles for GPP "...effective environmental, ethical, social and economic criteria will be promoted gradually and progressively in public procurement." It also establishes that in procurement processes that include

GPP criteria where the evaluation of offers is tied and all mechanisms for breaking the tie have failed, awarding should go to the proposal that inherently includes the most amounts of sustainable criteria (Autonomous City of Buenos, 2006).

- **Law N° 3871 for Mitigation and Adaption to Climate Change**: Prompted the City to design public policies establishing benchmarks and partial milestones in the issue (Autonomous City of Buenos Aires, 2011).

International Agreements

International treaties have offered a solid foundation for implementing GPP, Section 75 of the National Constitution elevates international treaties subscribed with other nations, international organizations and common markets to higher hierarchy than laws.

Argentina is also working towards the World Sustainable Development Agenda 2030 comprised of 17 Sustainable Development Objectives adopted by UN member countries. Public and private sector are participating in the process of adapting the 2030 agenda; Argentina's agenda will be defined by the end of 2017. (UN, 2017)

In addition, Argentina is a signatory to the following treaties and agreements regarding environmental protection:

- Paris Climate Accord (Law Nº 27270, 2016) (UN Climate Change, n.d.)
- Kyoto Protocol (Law Nº 25438 and Law Nº 27137, 2015) (UNFCCC, n.d.)
- United Nations Framework Convention on Climate Change (Law Nº 24295, 1994)
- Protocol on Environmental Protection to the Antarctic Treaty (Law Nº 24216, 1993) (Secretariat of the Antarctic Treaty, n.d.)
- Vienna Convention for the Protection of the Ozone Layer (Law Nº 23724, 1989) (UNEP, n.d.)
- Montreal Protocol on Substances that Deplete the Ozone Layer (Law Nº 25389, 2001) (UNEP, 2001)
- Biological Diversity Agreement (Law Nº 24375, 1994 and Decree Nº 1347/97) (CDB, n.d.)
- Framework agreement on the environment of MERCOSUR (Law Nº 25841, 2003) (ecolex, n.d.)
- Ramsar Convention on Wetlands of International Importance (Law Nº 23919, 1991) (Ramsar, n.d.)
- Convention Concerning the Protection of the World Cultural and Natural Heritage (Law Nº 21836, 1978) (WHC, 1972)

- United Nations Convention to Combat Desertification (Law N° 24701, 1994) (UNCCD, n.d.)
- Basel Convention on the Control of Transboundary Movements of Hazardous Wastes and Their Disposal (Law N° 23922, 1991 and Amendment Law N° 26664, 2011) (Basel, n.d.)

The City of Buenos Aires

The City of Buenos Aires is the Capital of Argentina and seat of the Federal Government. It is an autonomous territory and it is considered one of the 24 districts of the country. Argentina's socio- economic development has always been linked to the city as it consistently contributes a considerable portion of the country's GDP.

Given its high purchasing power, the GCBA is an important and influential actor in the goods and services market; with great potential to promote policies and create incentives towards sustainable development that generates benefits for the private sector and society as a whole.

Public Policy Potential of a Megacity

The Gross Geographic Product (GGP) of the Autonomous City of Buenos Aires comprises 20% of Argentina's total GDP. The City's per capita GGP is more than three times the average per capita GGP in the rest of the country. The City of Buenos Aires has grown along with the rest of Argentina in recent years and has exceeded the country's average in the growth rate of its Gross Regional Product (GRP), reaching more than 10% economic growth over the last 5 years. (GCBA, 2017)

Buenos Aires' economy is primarily comprised of the services sector, which represents 63% of real GGP and city nominal added value. Other sectors of note in GGP include manufacturing (15%), commerce (12%) and construction (10%) (GCBA, 2017).

The city has a population of approximately 3 million, however it borders with 40 municipalities that constitute the Greater Buenos Aires Metropolitan Area (GBAMA). Geographically it constitutes 1% of the country's territory and 37% of the population at 15 million (Argentina 2010 National Census). It is the fourth biggest metropolitan area in the Americas behind New York, USA, Sao Paulo, Brazil and Mexico DF, Mexico.

Given its status as a "megacity" and its impact on the country's development, a Consultative Commission for the Greater Buenos Aires Metropolitan was created. Its main objective is to articulate dialogue among jurisdictions in order to adopt innovative public policy for human, economic, technological, environmental and urban development (AMBA, 2016).

Environmental Efforts

The GCBA has taken on the "Zero Garbage" initiative that comprises a set of policies aimed to better manage urban solid residues' processing and disposal. The initiative also promotes sustainable and healthier means of transportation, increase of green areas, regeneration of native biodiversity, efficient use of natural resources and adoption of renewable energy sources. Some important milestones that the City of Buenos Aires has in its efforts for environmental sustainability are:

- In 2009, the 2010-2030 Climate Change Action Plan was issued with the objective of carrying-out a strategy for mitigating and adapting to climate change, and creating an inter-institutional task force. (UN, 2017)
- In 2015, the environmental protection agency of the GCBA issued the Climate Change Action Plan 2016-2020, although independent the plan is aligned to the nation's objectives for 2030. (GCBA, 2015)

The GCBA has also established an extensive joint agenda with other cities in the region in order to exchange knowledge and experiences to continuously improve climate change action plans. Some of the international forums where GCBA participates are:

- Iberoamerican Union of Capital Cities (UCCI, n.d.)
- C40 Cities Climates Leadership Group (n.d.); Carbon Disclosure Project (CDP, n.d.)
- Global Protocol for Community-Scale Greenhouse Gas Emission Inventories (GPC)
- Bonn Centre for Local Climate Action and Reporting (CARBONN, n.d.)

GPP Strategy in The City of Buenos Aires

The GPP program of the GCBA has the objective of increasing the number of suppliers that provide goods and services with sustainability standards and improving the procurement decisions of the purchasing agents of the system.

Besides the regulatory framework, some of the key operational elements of success for the implementation of GPP have been:

Public Procurement Governance

The GCBA has adopted centralized oversight and monitoring and decentralized operations as their choice for public procurement system, enabling a centralized and coordinated collection of information, planning, regulation, training, while monitoring the achievement of strategic objectives.

The General Procurement and Contracting Directorate (GPCD) based at the city Ministry of Finance is the public entity in charge of oversight and monitoring of the GCBA procurement and contracting system. Its main function is to promote, lead, articulate and monitor public procurement activities, ensuring that they are carried out in accordance with the principles of efficiency, efficacy, free competition, fair access, transparency, and sustainability (Buenos Aires Ciudad, n.d.).

Also, the regulatory framework establishes operational procurement decentralization for over 130 Procurement Operational Units (POUs) within the GCBA.

E-Government Procurement (E-GP) Key First Step for GPP

The GPCD has implemented a transactional electronic procurement system (e-GP), named Buenos Aires Compras (BAC) that has profoundly changed the way procurement is managed in the metropolitan region and is becoming the cornerstone

Figure 6. Screenshot of e-GP portal: Buenos Aires Compras (BAC)
Source: www.buenosairescompras.gob.ar

for the current federal e-GP system. This tool was crucial in order to operationalize GPP criteria in public purchasing.

Increased transparency has produced an improved dialogue with external and internal interest groups, regarding different initiatives for GPP. In turn, this dialogue has generated a good environment to further the implementation of GPP.

Increased availability of procurement information has enabled more suppliers to participate in the public procurement market. In turn, this enhanced engagement has made it easier for the GPCD to survey and gather information on the real capabilities of the private sector to supply goods and services under GPP criteria and fine-tune the pace and progressiveness of GPP implementation. At the same time, the data provided by the BAC system has enabled suppliers to better adjust their production strategies to qualify and benefit from the new GPP criteria.

An eGP platform has also facilitated the implementation of framework agreements where vendors do not have to supply goods or services upfront, instead several suppliers commit to provide goods or services over a period of time, better matching actual demand from POUs. Framework agreements are particularly important for GPP as they gather a high number of suppliers of similar characteristics under one process/contract that could potentially include GPP criteria. Modalities that aggregate public sector demand are popular among suppliers as it helps optimize their inventory management and manufacturing, this additional efficiency in the supply chain eventually reduces unnecessary use of raw materials, energy and natural resources for the production of goods.

The implementation of eGP at several stages of the procurement process also meant significant reductions in the use of paper for suppliers and the public sector. However, the most important contribution of eGP to the efforts of the GPCD and the GCBA is that it provides a platform where GPP recommendations and criteria can be easily deployed and monitored.

Inter-Institutional Engagement

Synergies have been pursued with different networks, organizations and external stakeholders in order to access relevant expertise, identify risks and fine-tune strategies, but most importantly to learn from previous experiences and not face reform in an isolated manner. Some of the key partnerships for GPP implementation in the GCBA have been:

- **Inter-Institutional Taskforce for Mitigation and Adaptation to Climate Change:** In 2014, the GPCD joined the taskforce putting public procurement at the forefront of the sustainability debate and strengthening the process to make public entities in the city efficient and sustainable buyers in matters related to the environment and energy consumption (Buenos Aires, n.d.).

- **Global Lead City Network on Sustainable Procurement (GLCN):** In 2016, Buenos Aires became the first Latin American city invited to join this network. The GLCN seeks to promote sustainable production and consumption through the implementation of GPP and innovation procurement. Associated cities act as ambassadors in their regions and establish ambitious and quantifiable objectives. Member cities include Seoul, Montreal, Denver, Rotterdam, Oslo and Buenos Aires (GLCNSP, n.d.).
- **Argentina Federal Public Procurement Network:** Launched in 2012, this network meets three times a year, bringing together subnational procurement entities and the federal procurement office. The GCBA represented by the GPCD participates actively in this network by leading a working group that gathers knowledge on GPP regulation and initiatives from the different jurisdictions in the network. As a result, GPP has been included as one of the main pillars of the Strategic Action Plan for the Network for the upcoming years.
- **Cooperation Agreement with Chile General Directorate of Public Procurement - ChileCompra:** Signed in 2013, the agreement is a GPCD initiative to promote technical cooperation in the area of GPP between both entities.
- **Inter-Institutional Sustainability Team at the GCBA:** Partnership between the two main entities at the GBCA in matters of GPP; the Environmental Protection Agency and the GPCD.

Flexible, Gradual, and Progressive Strategy

Since 2012, the GPCD has worked to include GPP criteria in the public procurement processes, using a flexible, gradual and progressive strategy that is reflected in activities such as:

- Awareness and training of public purchasers and vendors along the supply chain;
- Design of GPP directives and criteria for internal dissemination within the public administration and the country as a whole;
- Incorporation of sustainable criteria in bidding documents and evaluation
- Creation of synergies at the local, national and international level with public and private entities.

The strategy promotes responsible and smart purchasing by the POUs, incentivizing an innovative environmentally-friendly and participatory market with gradual and sustained growth.

The GPP strategy implemented by the GCBA aims to contribute to the following environmental goals:

- Reduce greenhouse gas emissions related to public procurement.
- Improve public procurement efficiency through the comprehensive analysis of costs and real demand.
- Enable more sustainable development without affecting levels of public expenditure.
- Promote corporate social responsibility.
- Adjust regulatory framework so that sustainable policies have clear operational rules that guarantee transparency and consistency over time.
- Establish an internal and external cooperation network that allows productive and agile exchange of information, experiences and capacity.

Figure 7. Screenshot of GPP portal of the GCBA
Source: www.buenosaires.gob.ar/hacienda/compras/compras-sustentables

- Train public officials to become exacting and discerning consumers
- Lead through example in environmental matters.

Training and Change Management

Providing information and training to POUs and suppliers has been an effective strategy to increase the number of vendors providing sustainable goods and services. Some of the key change management activities being carried-out include:

- Frequent consultation meetings with suppliers from priority sectors have been used to evaluate possible alternative solutions and present future GPP criteria. This has proven an effective mechanism to inform the private sector about the GPP initiative and gather valuable feedback from key sectors. Aside from traditional sectors engaged, meetings have been recently expanded to suppliers of marketing and merchandising, catering, and graphic arts, among others.
- Training and Awareness Workshops were implemented for over 130 operators of the eGP BAC system across the GCBA jurisdictions.
- GPP training for the GPCD staff was carried-out, including a series of internal workshops for the operational management offices within the GPCD. Workshops included training regarding the role of the GPCD in implementing GPP and supporting the POUs.
- A series of training sessions for POUs regarding many aspects of GPP implementation, including sessions on changes to the regulatory framework, the eGP platform BAC and new practices on GPP. Special awareness sessions were held with POUs at important line ministries such as Health and Education.
- Virtual training modules were implemented in the BAC and a virtual GPP introductory course was designed in partnership with the Public Administration School of the GCBA.
- A wide array of activities was carried out with media, academia, civil society to improve the GPCD informational portal and the eGP platform BAC.

Sustainable Criteria of the Buenos Aires GPP Program

Environmental and sustainability criteria can be introduced at many different stages of procurement, including when identifying demand for goods and services, when defining awarding criteria, when drafting technical specifications, at the launch of the bidding announcement, during the evaluation of offers, at the moment of awarding and even during the execution of the contract through monitoring and evaluation.

The strategy of the City of Buenos Aires was to provide tools to promote GPP in all stages of procurement. Some of these tools are:

GPP Guidelines and Recommendations

Guidelines were developed for POUs to provide GPP recommendations for the drafting of bidding documents, evaluating of offers and /or during the selection of suppliers. Some of these guidelines include:

- **Progressive Improvements of Sustainable Management of Packaging and Containers:** The guidelines include information, technical principles and recommendations to reduce volumes, proper disposal and improved quality of residues resulting from packaging goods procured by the GCBA.
- **Sustainable Use and Procurement of Paper:** Provides life-cost analysis guidelines for paper products, to be introduced in procurement and disposal decisions.
- **Recommendations for Energy Efficiency:** Supplies information and tools to properly implement life-cost analysis in procurement of energy-related goods and services. These guidelines will enable POUs to appraise not just acquisition and installation costs, but performance throughout lifecycle, capabilities to satisfy specific regional needs, durability, efficient use of resources and environmental impact (Buenos Aires, n.d.).

International and regional experiences on implementing GPP were taken into consideration while choosing topics for the recommendation guidelines, as well as the local and national regulatory frameworks, and consultation with stakeholders. The guidelines are available online and a first edition is being printed at the GCBA Printing Shop certified by the Forest Stewardship Council (FSC).

GPP Technical Specifications

Technical specifications have been implemented in many areas suitable for GPP implementation. A few examples include:

- Printing Paper offers should certify manufacturing through sugar cane processing or possess FSC or comparable certification.
- Office Furniture offers should include ergonomic information and accessibility features; also they should include information regarding plastics used.

- Packaging of goods procured should be manufactured with recycled materials or made with recyclable materials (paper, cardboard, polyethylene, polypropylene)

GPP Criteria and Voluntary Clause of Disclosure in Framework Agreements

The GPCD has incorporated extensive recommendations, technical criteria, information requests and verification mechanisms in the following framework agreements:

- Visual and Graphic Arts:
    - Registration in the "Environmental Impact and Residue Management Registry"
    - Minimum requirements for packaging
    - Information regarding the sustainability of paper, ink and other materials used
    - Information regarding sustainability of operational practices during preprinting and printing
- Television Sets:
    - Energy efficiency requirements
    - Recyclable packaging
- Toners and ink cartridges:
    - Responsible disposal management
    - Reverse logistics
    - Durability and quality of the product
- Scanners:
    - Energy efficiency certification
    - Digital manuals and documentation
    - Recyclable packaging
    - Disposal of packaging by supplier after installation
- Catering:
    - Voluntary Sustainability clauses regarding food, utensils, containers and packaging
    - Training for personnel
- Water:
    - Recyclable bottling and packaging

Many suppliers can be awarded framework agreements; a Voluntary Clause for Disclosure of Sustainability Information has been included for these suppliers. Although not a mandatory requirement, POUs can use the provided information to select the more convenient offer from a GPP perspective and issue purchasing orders independent of pricing.

PRELIMINARY RESULTS OF THE GPP PROGRAM

After 5 years of GPP implementation, a sufficient amount of data has been generated to start assessing the preliminary results of the GPP strategy of the City of Buenos Aires. The eGP system BAC will be the primary tool to measure GPP outputs for the GPCD, different modules are under development to monitor implementation, compliance and overall adoption of GPP.

The GPCD is undertaking substantial efforts to measure preliminary results that include the development of specific GPP indicators to be monitored and evaluated. It is not yet possible to infer outcomes related to the objectives of the Climate Change Action Plan 2020, as it goes beyond the GPCD mandate to monitor them. However, it is expected that implementation of the GPP strategy has already produced positive outcomes in energy efficiency and residue management; which will in turn will cause a reduction of greenhouse gas emissions.

Indicators to Monitor GPP Implementation

Indicators have been identified for an initial assessment of GPP implementation in the City of Buenos Aires. They have been divided in the following categories:

- Procurement processes that include GPP criteria:
    - Percentage of centralized procurement with GPP requirements
    - Percentage of centralized procurement with GPP recommendations
    - Percentage of total procurement with GPP requirements and recommendations
    - Percentage of POUs that use GPP recommendation guidelines
    - Percentage of public officials trained on GPP
    - Percentage of SMEs
- Expenditure in goods and services procured with GPP criteria:
    - Value and percentage of purchasing orders with energy efficiency requirements and/or recommendations
    - Value and percentage of purchasing orders with sustainable requirements and/or recommendations for packaging

- Value and percentage of purchasing orders with sustainability requirements and/or recommendations for paper use and consumption
- Value and percentage of purchasing orders with requirements of reduction and quality of residues generated
- Value and percentage of purchasing orders with requirements and/or recommendations for exclusion of toxic components

GPP Outputs Within Framework Agreements

During the implementation of the GPCD GPP strategy, framework agreements have proven to be a procurement modality well-suited for introducing GPP criteria. The flexibility provided by aggregating GCBA demand for goods and services allows the inclusion of mandatory and voluntary sustainability standards, in line with the flexible, gradual and progressive strategy outlined by the GPP implementation plan.

After 5 years since the GPP initiative started, output figures show successful implementation of GPP criteria and practices within GCBA framework agreements. Currently, more than half of the GCBA framework agreements include GPP requirements and/or recommendations. Expenditure on framework agreements with GPP requirements and/or recommendations surpasses USD$ 100 million and represents over 60% of the total.

Figure 8. Number of GCBA framework agreements that include GPP criteria
Source: GCBA (2017)

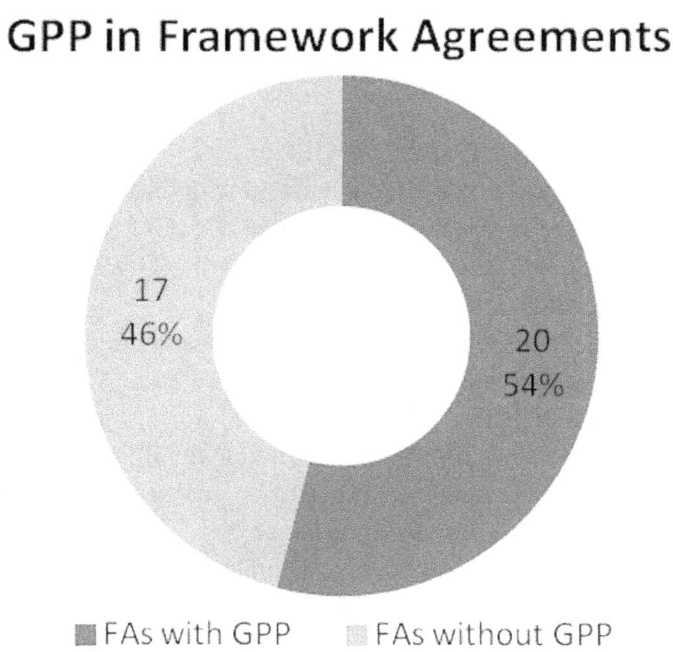

Table 3. Current framework agreements that incorporate GPP requirements and/ or recommendations

Goods	Services
1. Appliances, including air conditioning 2. Bookstore supplies and copiers 3. Bathroom supplies 4. Chemicals for pool maintenance 5. Cleaning supplies 6. Computers, desktops and laptops 7. Disposable diapers 8. Hardware, building materials and wood 9. Fire extinguishers 10. IT supplies 11. Lighting and electrical materials 12. Scanners 13. Televisions 14. Toner and ink cartridges 15. Water	1. Catering 2. Event Management 3. Merchandising 4. Painting 5. Visual and graphic arts

Source: GCBA (2017).

Figure 9. Expenditure in framework agreements that include GPP criteria
Source: GCBA (2017)

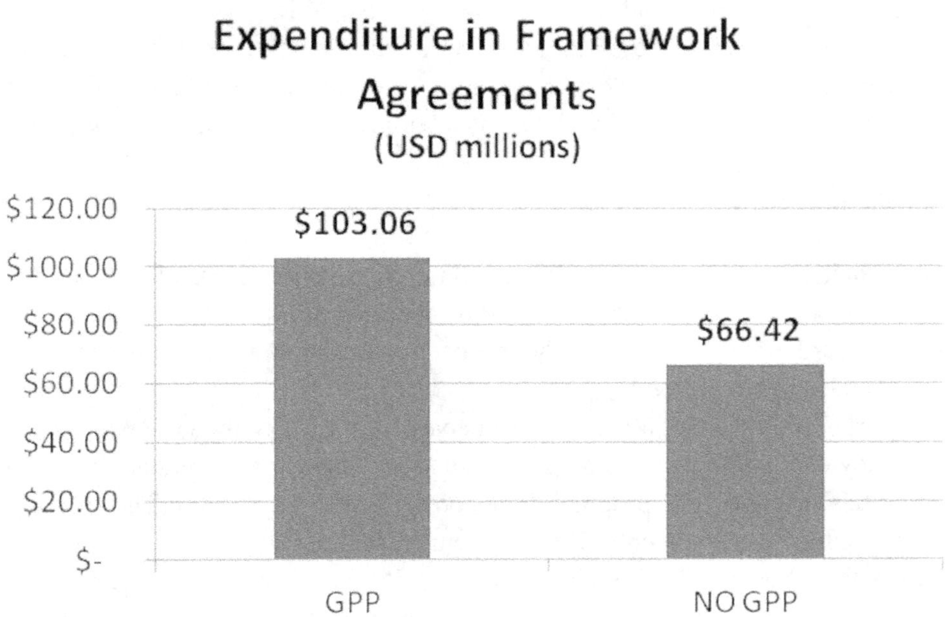

Lessons Learned

Since the inception of the GPP initiative in 2012, one main factor stands out as a key lesson learned for any jurisdiction that wishes to replicate the work done by the GCBA: The strategy that proved effective for GPP adoption is a gradual, flexible and progressive implementation with constant training, information and change management.

Some other specific lessons were learned during this process include:

Internal Engagement and Change Management

The GCBA faced significant risks by implementing GPP in a public administration focused on procuring goods/services based solely on the lowest economic offer at the moment of acquisition/provision. There was a possibility in the short-term, that POU's budgets could notably increase due to the selection of goods/services with GPP criteria; additionally, there was the chance of deserted procurement processes if not enough vendors presented offers, as they could be dissuaded by the new sustainability standards. Potential risks were amplified by public officials traditionally worried about administrative and legal scrutiny of their procurement decisions.

The solution to mitigate these risks was to carry-out extensive internal engagement at different levels of the GCBA. Creating awareness about the flexible and gradual strategy for implementing GPP was critical to resolve initial concerns by the POUs; as well as taking their feedback into account when deciding if certain GPP criteria should be voluntary or mandatory. Workshops were held for users of the BAC system and guidelines were published with technical information regarding the new policy.

Private Sector Outreach

It was challenging to introduce GPP criteria into a supplier market that has solely focused on winning government contracts by providing the lowest price offer. If new GPP standards were rolled-out without proper consultation it could disincentivize many vendors from presenting offers.

A series of consultation meetings were carried out where the GCBA presented the policy challenges faced and the private sector expressed potential issues and difficulties that the GPP program could present for their sectors. The meetings also served as valuable feedback on the actual capabilities of the supplier market to adapt their production of goods/services to GPP standards. Finally, the meetings were useful to establish that the public procurement system will not only consider the lowest price of acquisition but will take into consideration life-cycle costing, value for money and environmentally-friendly production processes.

Leveraging of Framework Agreements

The initial years of GPP implementation, coincided with a widespread adoption of framework agreements, which became a valuable tool to implement GPP, as this modality is characterized for increased dialogue and participation of suppliers during the preparatory stages of the agreement. The flexibility provided by this business model made it easy to incorporate mandatory and voluntary GGP provisions.

The agreements promoted participation of smaller but GPP-complaint suppliers, as they were not required to provide the full amount of goods/services stated in the agreement. A mix of mandatory and voluntary requirements was implemented and optimized depending of the type of contract, this prevented reduction of suppliers participating in framework agreements while giving POUs information to justify purchasing from more sustainable vendors options even if they didn't have the lowest price offer. Framework agreements provided valuable information regarding suppliers' capabilities to provide sustainable products and environmentally-friendly services. Finally, this procurement modality has supplied data that will support the introduction of GPP criteria in traditional procurement processes.

Introduction of Voluntary Disclosure of Information Clauses Related to GPP

During the first stages of implementation of the GPP program, mandatory requirements of sustainable or environmentally-friendly processes could discourage suppliers from presenting offers in the public procurement market.

Voluntary Disclosure of Information Clauses have allowed suppliers that have ecological certifications and/or sustainable production to promote their goods and services at the offering stage and provide the purchasers with information to select the most convenient offer even if it is not the lowest under traditional standards. This also provides with valuable data about how the market reacts to GPP which in prompts the introduction of more mandatory measures in the future.

The work carried-out by the GPCD in implementing GPP will be valuable as the entity will play an important role facilitating the public policy objectives mentioned above. The GPCD will be working coordinately with the BAYOGOC to identify procurement processes where GPP recommendations can be incorporated and provide support and advice to contracting entities for the Games.

CONCLUSION AND RECOMMENDATIONS

Conclusion

The City of Buenos Aires started its GPP initiative in 2012. Preliminary results after 5 years of gradual implementation show successful adoption of GPP practices within procurement operations and processes, especially within framework agreements. Currently, 20 framework agreements incorporate GPP recommendations and/or requirements; representing 60% of total expenditure and 54% of total number in framework agreements.

After analysis of the GPP strategy of the GCBA and its preliminary success, it could be concluded that the critical factors for its success include:

Gradual and Flexible Approach

The GPP strategy was progressively rolled-out mitigating risks associated with public administration adoption and supplier participation. Flexibility is also evidenced in measures such as the Voluntary Disclosure of Information Clauses; non-mandatory GPP recommendations; and guidelines for sustainability. Such measures played a critical role in the successful GPP program; they promoted participation from GPP-compliant suppliers without alienating the overall universe of vendors. Additionally, they created an environment for gradual introduction of mandatory GPP criteria.

Stakeholder Engagement

The GPCD carried out extensive consultation and awareness campaigns with a wide array of stakeholders such as public administration, private sector, and civil society. Additionally, the GCBA has partnered with different public and private institutions, international organizations, and procurement agencies from other countries to leverage best practices and experiences implementing GPP.

For a successful implementation of GPP it is necessary to properly fund and strategically plan outreach activities such as workshops, training sessions, virtual courses and consultation meetings. Building awareness among public purchasers is a constant challenge that should not be underestimated as it is crucial to enable permanent results. Furthermore, it is necessary to have an outreach strategy to engage suppliers and private sector organizations, in order to properly fine-tune and gradually roll-out new GPP initiatives that foster the capabilities of the supplier market without disincentivizing it.

Electronic Procurement System

The GCBA was in a privileged position to support public policies such as GPP, as they had a transactional eGP platform. The BAC system was a critical tool to disseminate information to public officials, incorporate GPP recommendations and requirements, and provide data to evaluate the initiative. An eGP transactional system expedites the implementation of GPP, as it streamlines incorporation of GPP criteria at different stages of the procurement processes. A transactional eGP system also enables introduction of requirements/recommendations in different procurement modalities.

Different Types of Procurement Models

The GPCD had different procurement modalities in place, which greatly facilitated the adoption of GPP practices in the City if Buenos Aires. Framework agreements were particularly effective as they grouped suppliers from the same industry making it possible to socialize GPP recommendations/requirements. Framework agreements, reverse auctions and centralized purchasing can play a critical role in the successful implementation of GPP; their inherent characteristics promote gradual introduction of GPP standards, enhanced dialogue with suppliers, and mitigate the risks associated with GPP adoption.

Recommendations

Preliminary results of GPP have favorable for the GCBA strategy. Some recommendations to keep furthering implementation and achieve the overarching policy objectives include:

1. **Furthering Monitoring and Data Analysis:** Extensive data analysis should be carried-out to evaluate implementation and adoption of GPP practices among the different stakeholders. Modules being developed for this purpose on the BAC will be key to undertake this effort.
2. **Assessment of Savings and Value for Money:** In partnership with relevant organizations and academia, methodologies to assess savings from the adoption of GPP processes could be developed. Information on potential savings can be critical to produce buy-in and support from decision makers in the public administration.

3. **Development of a Methodology to Evaluate Environmental Impacts:** As the ultimate purpose of the GPP program of the City of Buenos Aires is to protect the environment; it is necessary to start documenting and quantifying the role of GPP in reducing of greenhouse gas emissions and sustainable production practices.
4. **Expanding to Procurement of Works:** Finally, as part of the gradual approach that has successfully introduced GPP in the provision of goods and services, it should be considered to start developing a program to reproduce this strategy in the contracting of public works.

ACKNOWLEDGMENT

Marisa Siboldi, Sustainable Procurement Coordinator for the Government of the City of Buenos Aires served as collaborator for the development of this chapter, providing her expert advice and insights about the topic.

The authors would also like to thank the Undersecretary of Operational Management, Claudia Jaime and General Director of Procurement and Contracting, Marisa Tojo from the Government of the City of Buenos Aires for the valuable information and data provided for the chapter.

REFERENCES

AMBA. (2016). *Lanzamiento de la Comisión Consultiva Del Área Metropolitana.* Retrieved July, 2017 from: https://www.gba.gob.ar/medios/noticias/lanzamiento_de_la_comisi%C3%B3n_consultiva_del_%C3%A1rea_metropolitana

Argentina. (2010). *National Census Portal.* Retrieved July, 2017 from: http://www.indec.gov.ar/definitivos_bajarArchivoNacionales.asp?idc=8&arch=x&c=2010

Argentina General Environment Law 25.675. (2002). Retrieved July, 2017 from: http://ambiente.gob.ar/wp-content/uploads/POLITICA-AMBIENTAL-NACIONAL_Ley-25675.pdf

Argentina. (n.d.). *Argentina electrical emergency decree 134/15.* Retrieved July, 2017 from: https://www.boletinoficial.gob.ar/#!DetalleNormaBusquedaAvanzada/138328/20151217

Argentina Renewable Energy Law 26.190 amend. by Law 27.191. (n.d.). Retrieved July, 2017 from: http://servicios.infoleg.gob.ar/infolegInternet/anexos/250000-254999/253626/norma.htm

Autonomous City of Buenos Aires. (2006). *Aires procurement and contracting law 2095/06*. Retrieved July, 2017 from: http://www.buenosaires.gob.ar/areas/leg_tecnica/sin/normapop09.php?id=91501

Autonomous City of Buenos Aires. (2011). *Autonomous City of Buenos Aires Mitigation and Adaption to Climate Change Law 3871/11*. Retrieved July, 2017 from: http://www.buenosaires.gob.ar/areas/leg_tecnica/sin/normapop09.php?id=181526

Basel. (n.d.). *Basel convention on the control of transboundary movements of hazardous wastes and their disposal.* Retrieved from: http://www.basel.int/

Buenos Aires. (n.d.). Retrieved from: http://www.buenosaires.gob.ar/hacienda/compras/guia-para-el-consumo-y-compra-responsable-de-papel

Buenos Aires. (n.d.). *Climate change*. Retrieved from: http://www.buenosaires.gob.ar/areas/med_ambiente/apra/des_sust/pacc.php?menu_id=32408

Buenos Aires Ciudad. (n.d.). *Undersecretariat of operational management: Claudia Gabriela Jaime*. Retrieved from: http://www.buenosaires.gob.ar/hacienda/institucional-subsecretaria-de-gestion-operativa

C40. (n.d.). *C40 cities climates leadership group*. Retrieved from: http://www.c40.org/

Carbonn. (n.d.). Retrieved from: http://carbonn.org/

Casier, L., Huizenga, R., Perera, O., Ruete, M., & Turley, L. (2015) *Implementing Sustainable Public Procurement in Latin America and the Caribbean. Handbook for the Inter-American Network on Government Procurement (INGP).* International Institute for Sustainable Development (IISD). Retrieved May, 2017 from: https://www.iisd.org/sites/default/files/publications/iisd-handbook-ingp-en.pdf

CDB. (n.d.). *Biological Diversity Agreement*. Retrieved from: https://www.cbd.int/

CDP. (n.d.). Retrieved from: https://data.cdp.net/

Chartered Institute of Procurement and Supply (CIPS). (2017). *Procurement Glossary website*. Accessed June, 2017 at: https://www.cips.org/knowledge/procurement-glossary/q/

Commission of the European Communities. (2008). *Public Procurement for a Better Environment.* Retrieved May, 2017 from: http://eurlex.europa.eu/LexUriServ/LexUriServ.do?uri=COM:2008:0400:FIN:EN:PDF

Commission of the European Communities. (2016). *Benefits of GPP*. Accessed July, 2017 at: http://ec.europa.eu/environment/gpp/benefits_en.htm

Constitution of the Argentine Nation §41 and §43. (n.d.). Retrieved July, 2017 from: http://www.biblioteca.jus.gov.ar/argentina-constitution.pdf

Constitution of the City if Buenos Aires §8. (n.d.). Retrieved July, 2017 from: http://www.buenosaires.gob.ar/areas/leg_tecnica/sin/normapop09.php?id=26766&qu=c

Ecolex. (n.d.). *Framework agreement on the environment of MERCOSUR*. Retrieved from: http://www2.ecolex.org/server2neu.php/libcat/docs/TRE/Full/En/TRE-153663.pdf

EU Commission. (2016). *Commission of the European Communities. Eco-labels*. Accessed July, 2017 at: http://ec.europa.eu/environment/gpp/eco_labels.htm

General Services Administration (GSA). (2007). *Sustainable Facilities Tool*. Accessed July, 2017 at: https://sftool.gov/learn/about/41/sustainability

GLCNSP. (n.d.). *Global Lead City Network on Sustainable Procurement*. Retrieved from: http://www.glcn-on-sp.org/home/

Global Eco-labeling Network. (2012). *Environmental labels and declarations: How ISO standards help*. International Organization for Standardization (ISO). Retrieved May, 2017 from: https://www.iso.org/files/live/sites/isoorg/files/archive/pdf/en/environmental-labelling.pdf

Government of the City of Buenos Aires. (2015). *Climate Change Action Plan 2016-2020*. Retrieved July, 2017 from: http://cdn2.buenosaires.gob.ar/espaciopublico/apra/pacc_2020.pdf

Government of the City of Buenos Aires. (2017). *City of Buenos Aires Statistical System (SEC) (statistics portal)*. Accessed July, 2017 at: http://www.estadisticaciudad.gob.ar/eyc/

Government of the Commonwealth of Australia. (2013). *Department of Sustainability, Environment, Water, Population and Communities. Sustainable Procurement Guide*. Government of the Commonwealth of Australia. Retrieved June, 2017 from: https://www.environment.gov.au/system/files/resources/7b8df2bd-3bb9-49cc-b417-5f2eb6e0ce37/files/sustainable-procurement-guide.pdf

Grinier, S. (2014). *The INGP and the Latin-American Outlook on Sustainable Public Procurement.* INGP Annual Meeting. Retrieved May, 2017 from: http://ricg.org/download.php?len=es&id=293&nbre=1%20Presentation%20CPS%20RICG-2014%20%20Steven%20Griner%20Brasilia%20-%20Version%20Corta1.ppt&ti=application/vnd.ms-powerpoint&tc=Contenidos

International Institute for Sustainable Development (IISD). (2013). *Business and Sustainable Development: A Global Guide.* Retrieved June, 2017 from: https://www.iisd.org/business/tools/bt_green_pro.aspx

Jackson, P. (2012). *Value for money and international development: Deconstructing myths to promote a more constructive discussion.* OECD. Retrieved June, 2017 from: http://www.oecd.org/development/effectiveness/49652541.pdf

Marras. (2008). *The United Nations Sustainable Procurement Guide.* United Nations Environment Programme (UNEP). Retrieved July, 2017 from: http://www.greeningtheblue.org/sites/default/files/Buyingforabetterworld_0.pdf

Mendes, R. (2017). *Health in the green economy: Co-benefits to health of climate change mitigation.* World Health organization (WHO). Retrieved May, 2017 from: http://www.who.int/hia/green_economy/hgebrief_occ.pdf

Ministry of Energy. (n.d.). *Renewable energy law.* Retrieved from: http://www.energia.gob.ar/contenidos/verpagina.php?idpagina=3876

OECD. (2015). *Going green: Best practices for sustainable procurement.* Secretary-General of the OECD. Retrieved May, 2017 from: http://www.oecd.org/gov/ethics/Going_Green_Best_Practices_for_Sustainable_Procurement.pdf

Perera, O. (2015). *Implementing Sustainable Public Procurement in Latin America and the Caribbean. Handbook for the Inter-American Network on Government Procurement (INGP).* International Institute for Sustainable Development (IISD),. Retrieved May, 2017 from: https://www.iisd.org/pdf/2009/life_cycle_costing.pdf

Public-Private Infrastructure Advisory Facility (PPIAF). (2013). *Value-for-Money Analysis Practices and Challenges: How Governments Choose When to Use PPP to Deliver Public Infrastructure and Services* (Report from World Bank Global Round-Table). The World Bank Institute. Retrieved May, 2017 from: https://openknowledge.worldbank.org/bitstream/handle/10986/17622/840800WP0Box380ey0Analysis00PUBLIC0.pdf;sequence=1

Public Services and Procurement Canada (PSPC). (2006). *Policy on Green Procurement.* Government of Canada. Retrieved May, 2017 from: http://www.tpsgc-pwgsc.gc.ca/ecologisation-greening/achats-procurement/politique-policy-eng.html

Ramsar. (n.d.). *Ramsar convention on wetlands of international importance*. Retrieved from: http://www.ramsar.org/about-the-ramsar-convention

Secretariat of the Antarctic Treaty. (n.d.). *Protocol on Environmental Protection to the Antarctic*. Retrieved from: http://www.ats.aq/

SELA. (2015). *Public procurement as a tool for development in Latin America and the Caribbean*. Permanent Secretariat of SELA. Retrieved May, 2017 from: http://www.sela.org/media/268509/public-procurement-as-a-tool-for-development-in-lac.pdf

Sieb, A. (2011). *"Five benefits of green procurement" E-sourcing forum*. Determine Inc. Retrieved May, 2017 from: http://www.esourcingforum.com/archives/2011/11/29/five-benefits-of-green-procurement/

Strandberg, C., & Robinson, A. (2008). *Business Case & Benefits of Sustainability Purchasing*. The BuySmart Network. Retrieved May, 2017 from: http://www.buysmartbc.com/_Library/Resources/resource_bsn_business_case_to_sustainability_2008.pdf

The World Bank. (2013). Retrieved from: https://openknowledge.worldbank.org/bitstream/handle/10986/17622/840800WP0Box380ey0Analysis00PUBLIC0.pdf;sequence=1

UCCI. (n.d.). *Lo nuevo de la ucci*. Retrieved from: http://ciudadesiberoamericanas.org/

UN Climate Change. (n.d.). *Paris Climate Accord*. Retrieved from: http://newsroom.unfccc.int/paris-agreement/

UNCCD. (n.d.). *United Nations convention to combat desertification*. Retrieved from: http://www2.unccd.int/

UNEP. (2001). *Montreal Protocol on Substances that Deplete the Ozone Layer*. Retrieved from: http://ozone.unep.org/en/treaties-and-decisions/montreal-protocol-substances-deplete-ozone-layer

UNEP. (n.d.). *Vienna convention for the protection of the ozone layer*. Retrieved from: http://ozone.unep.org/en/treaties-and-decisions/vienna-convention-protection-ozone-layer

UNFCCC. (n.d.). *Kyoto protocol – targets for the first commitment period*. Retrieved from: http://unfccc.int/kyoto_protocol/items/2830.php

United Nations. (2017). *The sustainable development agenda 2030*. Retrieved from: http://www.un.org/sustainabledevelopment/development-agenda/

WHC. (1972). *Convention concerning the protection of the world cultural and natural heritage*. Retrieved from: http://whc.unesco.org/en/conventiontext/

Wikipedia. (2017). *List of Metropolitan Areas in the Americas.* Accessed July, 2017 at: https://en.wikipedia.org/wiki/List_of_metropolitan_areas_in_the_Americas

Williams-Elegbe, S. (2017). *Public Procurement and Multilateral Development Banks: Law, Practice and Problems*. Portland, OR: Hart Publishing.

Yaker, F. (2013). *Sustainable Public Procurement: A Global Review Final Report.* United Nations Environment Programme (UNEP). Retrieved May, 2017 from: https://www.globalecolabelling.net/assets/Documents/unep-spp-report.pdf

Chapter 7
Green Public Procurement Initiatives in East African Community (EAC):
The Way Forward

Rajesh Kumar Shakya
The World Bank, USA

ABSTRACT

This chapter focuses on the green public procurement initiatives taken by the countries in the East Africa. The East African Community (EAC) Head of States signed the East African Monetary Protocol Union (EMPU) protocol on November 30, 2013. One of the key ingredients in having a sustainable monetary union is to harmonize public financial management systems amongst the EAC Partner States and the EAMU protocol has referred the provision of harmonization of public procurement policies. The first stakeholder' workshop on the harmonization of public financial management standards in line with the EAC Monetary Union Protocol held during March 14-17, 2014 in Nairobi, Kenya identified green public procurement (GPP) strategy as one of the key policy areas requiring development and harmonization across the member states. This initiative should also be viewed in the context of the Member States' Public Procurement Reform Initiatives, which contains a suite of measures in relation to public procurement. Reform of the public procurement function is, and remains, driven by the need to obtain maximum value for public money in procuring works, supplies, and services.

DOI: 10.4018/978-1-5225-7083-7.ch007

INTRODUCTION

East African Community comprises of five countries: Burundi, Kenya, Rwanda, Tanzania and Uganda (Figure 1). The Environment and Development conference at Rio de Janeiro in 1992 was a significant milestone in building awareness about the green public procurement. In many African countries already, the export of products is threatened by international green trade barriers which directly affect business competitiveness in the global market. Besides, facing to a variety of competitive pressure from the external environment, the enterprises in the Africa region has initiated the green procurement to improve its competitiveness across whole supply chain from the procurement of raw materials, processing, production to delivery to enhance its competitiveness by minimizing the overuse and waste of materials, and recyclable packaging of the products.

Green procurement is becoming a cornerstone of environmental policies all over the world, especially popular at European Union Member State (Tukker et al., 2008). However, even in Europe, most of the organization have not fully implemented GPP. United Nations's Sustainable Development Goals (SDG) has played a driving role in making all emerging countries as well as developed countries to come forward with the sustainable and green initiatives to meet the SDGs.

East African Community (EAC) and its member states also have started Green Public Procurement (GPP) initiative to catch-up with the rest of the world. The initiative for green procurement by the East African Community (EAC) is in support of the goal of the United Nations (UN), which is to become climate neutral and environmentally sustainable gradually. In the case of public procurement, the inclusion of environmental, economic and social criteria in the procurement of goods, works, and by procuring entities help facilitate sustainable procurement. In this context, the East African Monetary Protocol Union (EMPU) protocol signed by the Head of States of the East African Community (EAC) on November 30, 2013, is the key milestone. One of the vital components of the protocol is having a sustainable Monetary Union to harmonize public financial management systems amongst the EAC Partner States. Alongside, the EAMU protocol has referred to the provision of harmonization of public procurement policies. The first stakeholder' workshop on the harmonization of public financial management standards in line with the EAC Monetary Union Protocol was held during March 14-17, 2014 in Nairobi, Kenya, which identified Green Public Procurement (GPP) Strategy as one of the critical policy areas requiring development and harmonization across the member states. In addition to that, 10th East African Procurement Forum held in Kampala, Uganda during 1-3 November 2017 also emphasized the need for harmonized green procurement guidelines in the EAC member states (PPF, 2017).

Figure 1. East African community
Source: Mapsofworld, 2018

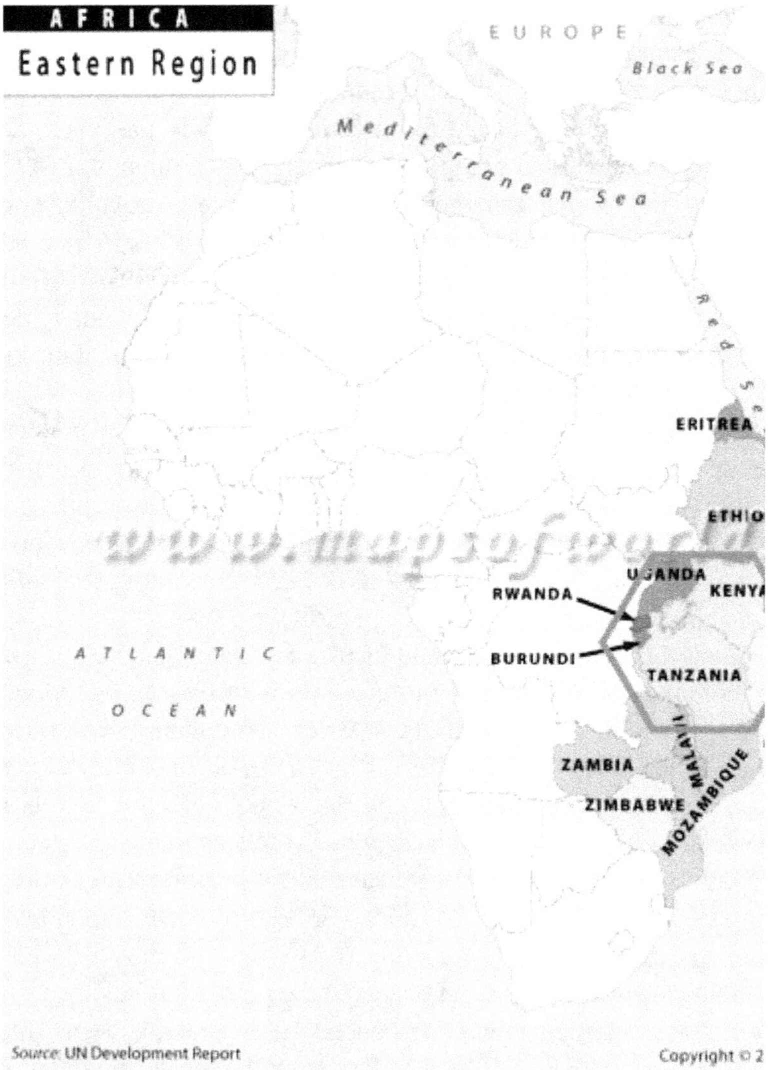

Public procurement is the tool for implementing development programs spending public fund. This GPP initiative should be viewed in the context of the EAC's recently announced harmonization policy plans, which contains a suite of measures concerning public procurement. Harmonization of the public procurement functions across the EAC states is primarily driven by the objective of achieving the best value for money in spending public fund for the use in development works and acquiring services. Another revolutionary agenda to harmonize the green procurement strategy is, therefore, a significant initiative, not just in introducing a sustainable development mindset within public procurement practices, but also in ensuring that EAC taxpayers' money is spent wisely. It is necessary to underline, that the annual public-sector procurement budget accounts for about 20% of EAC's GDP. This provides EAC's public sector with considerable leverage to stimulate the marketplace in favor of the provision of more resource-efficient, less polluting goods, services, and works.

STUDY ON GREEN PROCUREMENT POLICY IN AFRICA

Many researchers and projects have carried out some research and studies in Africa on green procurement policy. Many researchers and projects have carried out some research and studies in Africa on green procurement policy. AfDB (2014) wrote in its findings of a literature review on Sustainable Public Procurement (SPP) and Green Public Procurement (GPP) the minimum conditions must be met regarding awareness of environmental and social problems, a sense urgency, political will, and country ownership for starting the SPP/GPP. For SPP to thrive, the necessary policy and legal framework must be put in place, capacity must be built, and the demand and supply sides must understand the new requirements and the applicable techniques (AfDB, 2014). However, the EAC states had shown their commitment towards the green movements during the World Conference on Sustainable Development in 2002. Many a time, green procurement policies were tried to be tested at different levels of government and sectors. Some of the sectors, like energy and housing, have shown some good results in some specific projects. While certain government agencies seem to have moved ahead in developing green procurement guidelines and standards, the implementation of these standards again appears to be less than complete. This suggests that the process of preparation and execution of green procurement criteria has not been effectively rolled out across the procurement processes. Many of the governments, although having committed publicly to the development of green procurement policy, drive the work, and the program was placed on hold (Roose & De La Harpe, 2008). Green procurement is currently mainly being driven by environmental departments or energy departments without much involvement or contribution from the government policy in general (Roose & De La Harpe, 2008).

In Africa, most of the green initiatives are carried out at the initiation of enterprises and in silos. One such study was done at the Kenya pipeline corporation on determinants, which revealed that the green procurement could be successful only through the incentives to the procuring entities, suppliers, bureaucrats, and policymakers, needs unconditional pressure to implement, and need organization level readiness and capacity to understand and use the green procurement criteria (Bjorklund, 2011; Testa et al., 2012). It was also revealed that the supply market was not ready to produce green products and fulfil green criteria across the supply chain. The results indicate that domestic readiness is significant in the adoption of the green procurement rather than of focusing on the external market. In the meantime, many private enterprises in Kenya are working to improve the environmental performance of their operations and productions, and green procurement will be the inherent to this effort.

The private sector has been adopting green procurement for the last two decades for specific products like office paper, cleaning agents, renewable energy, and paints. Large enterprises have developed their green procurement policies for a broader range of goods and services. So far the business gets benefit out of these initiatives, green procurement may continue to be used in private enterprises.

ENVIRONMENTAL CRITERIA UNDER EAC MEMBER COUNTRY LAW

Despite the growing number of countries that have adopted GPP policies, GPP in Africa is still in its infancy (Akenroye, Oyegoke, & Eyo, 2013). According to Akenroye et al. (2013), African countries do not only lack the institutional and legal framework to implement GPP policies, but also political commitment.

Procurement laws in the EAC member states provide provisions of sustainable procurement, which can provide the opportunity for the procuring entities to incorporate environmental requirements into the procurement process. What is more, being a member of the EAC Member States should agree to subject its procurement procedures practices to a certain level of international and regional regulation, which recognizes standard rules and criteria, which must be seen in the award of the public contracts. The EAC can play a significant role in developing and adopting the policy framework which guides and assists the Member States in implementing procurement policies that seek to address not only economic but also environmental objectives. The purpose of coordinating at the EAC level, the procedures for the award of procurement contracts is to minimize barriers to the freedom of exchange and trade of goods and services across the member states. Therefore, to protect the interests of suppliers from the member states who wish to do business across the border, environmental policy should cover all the areas with the environmental impact to contribute to the preservation, protection, and improvement of the quality of the environmental problems not only in a country but across all the member states. It will foster competition and efficiency, and also seek to address environmental goals across the member states. Hence, It is highly appreciable that the recent amendments in public procurement laws and new public procurement policy documents in member states of EAC reflect the aspirations for the sustainable development by protection and improvement of the quality of the environment.

The following section looks at the country status of each of the EAC member states regarding incorporation of the sustainable or green public procurement provisions.

Kenya

Kenya is ranked at 130 position out of 180 countries in 2018 Environment Performance Index (EPI, 2018) assessed on 24 performance indicators across ten issues categories covering environmental health and ecosystem vitality.

Kenya does not have any Green Public Procurement (GPP) strategy. The new Public Procurement and Assets Disposal Law enacted in December 2015 provides Article 3 (i) one of the guiding principles of procurement as "promotion of local industry, sustainable development, and protection of the environment." On disposal methods, article 165 (1) (e) provisioned waste disposal management and 165(2) states, "Despite subsection (1) or any other provisions of this Act, radioactive or electronic waste shall be disposed of only to persons licensed to handle the respective waste under section 88 of the Environmental Management and Co-ordination Act, 1999.

A task force is formed to develop procurement regulation, which is supposed to prepare regulation related to GPP in addition to other regulations.

The National Environment Management Authority (NEMA) was established under the Environmental Management and Co-ordination Act No. 8 of 1999 (EMCA) as the principal instrument of Government for the implementation of all policies relating to the environment. EMCA 1999 was enacted against a backdrop of 78 sectoral laws dealing with various components of the environment, the deteriorating state of Kenya's environment, as well as increasing social and economic inequalities, the combined effect of which negatively impacted on the environment. The supreme objective underlying the enactment of EMCA 1999 was to bring harmony in the management of the country's environment (NEMA, 2018). In the case of green procurement, NEMA is still not mandated any particular role.

Uganda

Uganda is ranked at 145 position out of 180 countries in 2018 Environment Performance Index (EPI, 2018) assessed on 24 performance indicators across ten issues categories covering environmental health and ecosystem vitality.

Uganda does not have any Green Public Procurement (GPP) strategy. The New Public Procurement and Assets Disposal Law enacted in 2014, Article 25 (2)(e) talks about the Statement of Requirement for Supplies, a functional description of the qualities, including any environmental or safety features required of the subject of the procurement." On disposal methods, article 15 (a) considers, "where national security, public interest, health and safety issues, legal and human rights issues or environmental considerations will be served if the public asset is converted or classified into another form." In practice, the provisions of the law are still not observed the enforcement of these provisions.

Additionally, Sustainable Public Procurement is being focused on the new National Public Procurement Policy Draft stressing on minimum damage in the environment and elimination of waste.

The National Environment Act (NEA), Cap. 153, stipulates the Mandate of National Environment Management Authority (NEMA) as the principal Agency in Uganda responsible for the management of the environment by coordinating, monitoring, regulating, and supervising all activities in the field of environment, but not a particular role given in procurement process yet.

Rwanda

Rwanda is ranked at 148 position out of 180 countries in 2018 Environment Performance Index (EPI, 2018) assessed on 24 performance indicators across ten issues categories covering environmental health and ecosystem vitality.

Sustainable Public Procurement is not included in Rwanda Public Procurement Authority Law. There is a similar law titled, "Organic Law on protection, conservation and promotion of Environment in Rwanda 2005". The Republic of Rwanda has got a comprehensive Green Growth Strategy, under which all organizations and their activities have to comply. There is an existence of the Rwanda Environment Management Authority (REMA), but not a particular role is given in the procurement process yet. The Rwanda Natural Resources Authority under the Ministry of Natural Resources carries out environmental impact assessments in case of large infrastructure projects. There is no specific mention of Green Procurement compliance requirements in Standard Bidding Documents.

Tanzania

Tanzania is ranked at 119 position out of 180 countries in 2018 Environment Performance Index (EPI, 2018) assessed on 24 performance indicators across ten issues categories covering environmental health and ecosystem vitality.

Sustainable Public Procurement is given focus on the new National Public Procurement Policy Draft stressing on minimum damage in the environment and elimination of waste and establishment of the sustainable public procurement system.

A new bill on public procurement is underway, which is expected to cover the Green Public Procurement. General Procurement Service Authority (GPSA) and Assets management units do not consider green requirements in their procurement and programs yet. Public Asset Policy Framework is being prepared, and green requirement issues in case of disposal will be considered.

The National Environment Management Council (NEMC) came into being in 1983 when the Government of Tanzania enacted the National Environment Management Act No. 19 of 1983. NEMC was established with a broad mandate in response to the national need for such an institution to oversee environmental management issues and also implement the resolutions of the Stockholm conference (1972), which called upon all nations to establish and strengthen national environmental Councils to advise governments and the international community on environmental issues (NEMC, 2018). The National Environment Management Council (NEMC) is given any particular role n procurement process yet.

Burundi

Kenya is ranked at the last position (180) position out of 180 countries in 2018 Environment Performance Index (EPI, 2018) assessed on 24 performance indicators across ten issues categories covering environmental health and ecosystem vitality. This indicates the critical policy intervention requirement in Burundi to enhance the environmental situation.

Public Procurement Code of Burundi is the Act 1/01 issued on fourth. February 2008. Article 82 of the public procurement code talks about the tenderer selection criteria to comply with environmental standards among other criteria, but in practice, such criteria are not still practiced. Burundi's Poverty Reduction Strategy (PRS) for 2006-09 had a mention about the environment as one of the crosscutting issues to be addressed.

However, although the provisions on environmental matters were included in the law and procurement strategies, these are almost all still to be incorporated in the regulations, and their application both in law and practice is all to be formulated by the Member States or individual procuring entities. For the introduction of Green Public Procurement (GPP), it is recommended to start with the provision by defining EAC value thresholds of the procurement to make applicable to such procurement procedures instead of applying in all value procurements. Usually, it is recommended to apply the GPP criteria on the high-value contracts excluding low-value contract for the time being.

In the cases when the value of the contracts is below the thresholds indicated in the EAC threshold levels, national procurement rules have to fill the gap. Rules to be prepared by EAC and the EAC Member States should be published in the Portal of the EAC and the Central Public Procurement portal of each EAC Member States and should be updated every year. The Public Procurement Regulations of the each EAC Member States should require observing specific rules (for example, requirements for technical specifications, verifying Suppliers' qualification, evaluation of bids) while preparing the individual rules for public procurement. The procuring entities should be allowed to decide what kind of procedures to use when awarding these contracts. All in all, all the national rules, both the Public Procurement Law and the Regulations, have to observe the fundamental principles of the Green Public Procurement (GPP).

A procuring entity can include horizontal procurement criteria promoting environmental policy objectives, even if these criteria do not contribute to the direct economic advantage of the procuring entity. The criteria are linked to the subject-matter of the contract, comply with the general requirements of equal treatment, transparency and proportionality, are expressly mentioned in the contract documents or the Invitation to Bid and do not confer an unrestricted freedom of choice on the authority. Besides, the procuring entities should be allowed to define the criteria for determining the most economically advantageous bid and assigning the coefficient of such criteria.

Regulating public procurement in their respective Member States provide many opportunities for the procuring entities to include environmental considerations into the procurement procedures. Namely, if a procuring entity wishes to include environmental requirements into the procurement process, it can do so by, for example, incorporating such requirements into the technical specifications, by using the eco-labels, excluding on the ground of non-compliance with environmental obligations or by setting up environmental conditions for the performance of a contract. In such a way, primary procurement objectives can be balanced with the protection of the environment. However, the environmental provisions are almost all still voluntary. Therefore, the extent and frequency to which these provisions will be implemented depend on the governments of the Member States or individual procuring entities.

GPP MONITORING AND IMPLEMENTATION

The monitoring and implementation of Green Public Procurement will be critical to its integration into public procurement policies and practices across all sectors in the EAC Member States. Its successful implementation will ensure a consistent approach to GPP across the various sectors. While Green Public Procurement sets out a series of actions under the selected priority areas, individual sectors are likely to require the development of more detailed implementation plans to meet the requirements of their specific sectors. It is essential that a coordinated, partnership approach involving the EAC Member States be intensified to achieve effective implementation of the GPP Strategy.

To this end, the EAC in consultation with the Public Procurement Authorities and other key stakeholders will maintain a GPP Strategy Implementation Group, whose role will be to monitor and oversee the rollout of the GPP and to report on progress on a yearly basis. The Implementation Group's first annual progress report will include a detailed implementation plan for the sectors prioritized in this document, to include a timeline for each of the individual actions. The Implementation Group will also develop arrangements for ongoing consultation with relevant private sector stakeholders as appropriate. The Action Plan itself will be renewed on a five-yearly basis.

Performance Indicators

Green public procurement is an iterative process requiring on-going engagement with stakeholders. The range and relative cost of public procurement are continuously changing. Public procurers should keep themselves informed about changing technologies and processes, so that, where these are cost-effective and fit for purpose, relevant developments and innovations can be included in the procurement process.

There is a need to establish better performance relative to two GPP indicators, widely used to measure GPP at EAC level. These two indicators are the value of public procurement tendering processes/contracts with GPP criteria, relative to the overall value of public procurement contracts, and the number of such bidding processes/contracts relative to the total number. These two indicators are essential in order to monitor, review, evaluate and revise the embedding of GPP in each Member State. They also provide a "baseline scenario" which would also be useful for identifying existing products and services, evaluating current demand and monitoring this demand over time as actions are implemented.

In the future, evaluation at EAC level is likely to include the impact of GPP on carbon emissions and the impact of GPP on procurement costs, particularly on a life-cycle basis.

TRAINING AND EDUCATION

Proper training in GPP is essential to achieve an environmentally-beneficial outcome, such as training drivers to be more fuel-efficient and training cleaning staff about appropriate products and dosages. However, continuous education and training of all key procurement personnel are essential for this strategy to achieve its goals and for GPP to achieve its potential. This training effort should primarily focus on:

- Policymakers
- Managers
- Legal, financial and procurement officers

It is important to recognize that sustainable procurement and GPP procurements are increasingly being mainstreamed into a wide range of existing education and training initiatives. Universities offer a Masters degree in Strategic Procurement, which includes sustainable procurement and GPP issues. The Civil Service Training and Development Centers also can deliver training on Strategic Objectives in Public Procurement, which highlights GPP policy objectives. These developments should be facilitated and encouraged as much as possible and progress made should be noted in the reviews of GPP implementation carried out by the GPP Implementation Group.

FUTURE ACTIONS

Future actions should be developed within a qualitative context where possible. While adherence to the GPP processes will be a vital part of the further evolution of EAC's GPP policy, in practice each Member State will continue to define its targets in every sector to contribute to the overall 50% target. Member States should utilize the flexibility and dynamism inherent in this process, in order both to respect these broader objectives and obligations, while also maximizing value for money, innovation potential and market opportunities.

Overarching Key Actions

1. To ensure that GPP is successfully embedded in public procurement policies and practices across all sectors, the following key actions are proposed to be undertaken:

2. Government Ministries, Departments and Agencies (MDAs) with a procurement role shall progressively integrate green criteria into public sector bidding processes, as appropriate.
3. A GPP Strategy Implementation Group, comprising relevant Government MDAs, shall be tasked with reviewing the implementation of GPP on an annual basis, by reference to existing and future national and EAC targets.
4. The GPP Strategy Implementation Group's first annual progress report will include a detailed implementation plan for the selected priority areas, to include a timeline for each of the individual actions.
5. The GPP Strategy Implementation Group shall draw up terms of reference for further on-going research into GPP methodologies, target-setting, and effective implementation, evaluation, and monitoring.
6. The GPP Strategy Implementation Group's annual review should also report on the level of GPP training for procuring entities that have taken place in the year under review.
7. GPP content and reference to Green procurement should be integrated into all public procurement-related training provided to public entities. This should apply irrespective of whether the training intervention is developed and delivered "in-house" or by external providers.
8. In the case of the priority product groups, the Implementation Group's review will be based on available statistics and data for two key indicators – the number of contracts with GPP criteria relative to the overall number of contracts; and the value of such contracts relative to the overall public procurement spend.
9. Procuring entities, when addressing procurement capability in their organizational plans, should indicate a commitment to whole-of-government objectives including GPP.
10. Organizational plans should set out how any GPP skills gap will be identified and dealt with through education, training, awareness raising and dissemination of information. Those bodies with a significant procurement function should include these actions in their Training and Development Strategy/ Plan.
11. Procuring entities should ensure that procurement guidance material made available to staff (whether in hardcopy or electronic format) highlights the need to incorporate GPP criteria where appropriate into the tender and contract documents.
12. Where applicable and relevant to the subject matter of the contract, procuring entities shall specify environmental and energy efficiency criteria, among criteria specified for contracts being awarded by Most Economically Advantageous Tender (MEAT).

CONCLUSION

The world is talking a lot about sustainability issues, especially on the environment. Public authorities spend a significant amount of money every year on acquiring products, services or works which they need for the implementation of their functions. It is already widely recognized that public authorities have to implement not only functional but also social and horizontal objectives, such as the protection of the environment. Therefore, the procuring entities should contribute to the improvement of the environment and thus be a part of the sustainable solution by making environmentally friendly procurement decisions. The existing situation can be improved, and the primary procurement objective (the acquisition of goods, services, and works necessary to accomplish government mission in a timely, economical and efficient manner) can be balanced with the societal goal of protecting the environment by including environmental considerations into the different stages of procurement.

The Law on Public Procurement of the EAC Member States provides that the purchasing authorities can award a procurement contract by either the criterion of the lowest price or the criterion of the most economically advantageous tender (MEAT). Purchases based solely on the lowest price criterion often result in significant negative externalities, including environmental degradation. The criteria of the Most Economically Advantageous Tender, on the contrary, induces the inclusion of the environmental considerations into the procurement process and therefore fosters the sustainable consumption and production. However, public authorities rarely apply MEAT criterion because it is considered to be more involved and time-consuming. Furthermore, the procuring entities indicate that it is difficult to define the objective criteria for the evaluation of tenders. Consequently, the use of the lowest price criterion as a significant criterion for awarding the public contracts results in decreasing number and value of purchases, which include environmental requirements into the bidding process. This, in turn, stimulates the acquisition of products with a negative environmental impact.

REFERENCES

AfDB. (2014). *Summary of Literature on Sustainable Public Procurement/Green Public Procurement.* African Development Bank. Available at https://www.afdb.org/fileadmin/uploads/afdb/Documents/Procurement/Project-related-Procurement/Summary_of_Literature_on_Sustainable_-_Green_Public_Procurement.pdf

Akenroye, T., Oyegoke, A., & Eyo, A. (2013). Development of a framework for the implementation of green public procurement in Nigeria. *International Journal of Procurement Management*, *6*(1), 1–23. doi:10.1504/IJPM.2013.050607

Bjorklund, M. (2011). Influence from the business environment on environmental purchasing: Drivers and hinders of purchasing green transportation services. *Journal of Purchasing and Supply Management*, *17*(1), 11–22. doi:10.1016/j.pursup.2010.04.002

EPI. (2018). *Environmental Performance Index Report*. Yale Center for Environmental Law & Policy, Yale University Center for International Earth Science Information Network, Columbia University. Available at: https://epi.envirocenter.yale.edu/downloads/epi2018policymakerssummaryv01.pdf

NEMA. (2018). *National Environment Management Authority*. Available at https://www.nema.go.ke/index.php?option=com_content&view=article&id=1&Itemid=136

NEMC. (2018). *National Environment Management Council*. Available at https://www.unenvironment.org/nairobiconvention/national-environment-management-council-nemc-tanzania

PPF. (2017). *Report on the 10th East African Procurement Forum*. PPADB, Kampala, Uganda.

Roose, R., & de la Harpe, S. (2008). Good governance in public procurement: a South African case study. *PER: Potchefstroomse Elektroniese Regsblad, 11*(2). Available at http://www.scielo.org.za/scielo.php7pidM1727-37812008000200006&script-sciarttext

Testa, F., Iraldo, F., Frey, M., & Daddi, T. (2012). What factors influence the uptake of GPP (green public procurement) practices? New evidence from an Italian survey. *Econological Economics, 82*(2012), 88-96. Available at http://ec.europa.eu/environment/gpp/pdf/uptake_GPP_practices.pdf

Tukker, A., Emmert, S., Charter, M., Vezzoli, C., Sto, E., Andersen, M. M., ... Lahlou, S. (2008). Fostering change to sustainable consumption and production: An evidence based view. *Journal of Cleaner Production*, *16*(11), 1218–1225. doi:10.1016/j.jclepro.2007.08.015

Chapter 8
Green Public Procurement in Bhutan:
Success Story from Asia

Rajesh Kumar Shakya
The World Bank, USA

Pem Lama
GPP Bhutan, Bhutan

ABSTRACT

Bhutan is a developing country in Asia with a strong commitment to environment and sustainability. While most of the developing world is still trying to explore the opportunities to formulate the Sustainable Public Procurement (SPP) policies and legislation, Bhutan has successfully implemented the Green Public Procurement in Bhutan (GPP Bhutan) project. The three-year project started in 2014 and ended in June 2017. It has conducted research, trained public procurers, sensitized suppliers, and piloted green public procurement projects to introduce and implement green procurement in the country. This chapter aims to discuss how the government of Bhutan has embraced the value of green public procurement (GPP) and is setting an example for other countries of how to leverage the power of the public purse to advance sustainable development. The chapter presents the success story of Bhutan from the perspective of its implementation approaches. The Bhutan strategy encompasses all spheres of national policies in line with the sustainable or green public procurement (GPP)—whereby government authorities seek to procure goods, services, and works with a reduced environmental impact throughout their life cycle, as well as with an increased positive social impact—that have the potential to drive sustainable growth in Bhutan. The chapter also at places points to Bhutan's various national policies and strategies and illustrates some examples and experiences that other emerging countries can learn from.

DOI: 10.4018/978-1-5225-7083-7.ch008

INTRODUCTION

High up in the Eastern Himalayas is one of the greenest countries in the world. While many nations are struggling to reduce their carbon emissions, the Kingdom of Bhutan is already carbon negative: it takes more greenhouse gasses from the atmosphere than it emits. (Tutton & Scott, 2018)

CNN published an article by Mark Tutton and Katy Scott about Bhutan green achievement on October 11, 2018. It would be relevant to present an excerpt from the article here (shown in Box 1).

Bhutan has demonstrated how strategy plays a role in making the vision a success, that is also in the complex domain of Green Public Procurement. Bhutan was ideal for the implementation of the GPP project as green initiatives are not new at all in the country. Bhutan has taken a unique approach its development called Gross National Happiness (GNH), which seeks to achieve equitable and sustainable economic development while protecting its environment, preserving its culture and ensuring good governance. For decades now improving the Gross National Happiness (GNH), not the GDP, has been at the heart of all of Bhutan's political and economic agendas (Forest, 2017).

Box 1.

Sandwiched between China and India, Bhutan spans approximately 14,800 square miles -- roughly the size of Maryland. Its vast woodlands cover approximately 70% of the country and act as a natural carbon sink, absorbing carbon dioxide.

As a result, according to its own figures, this nation of around 750,000 people removes nearly three times as much CO2 as it produces.

Bhutan's ability to be a net carbon sink is partly down to its natural forests and the fact that it is relatively undeveloped -- most people work in agriculture or forestry -- which means it emits less than 2.5 million tons of CO2 each year. Luxembourg, for example, with a smaller population, emits four times as much.

There are other carbon negative countries in the world, and they are also heavily forested and undeveloped. But in Bhutan, there are other factors at play.

For the past 46 years the Bhutanese government has opted to measure progress not through its Gross Domestic Product, but through "Gross National Happiness," which places great emphasis on the protection of the country's rich natural environment.

"Bhutan is the only country in the world that by its own constitution protects its forests," explains Juergen Nagler, of the UN Development Program in Bhutan.

Environmental protection is enshrined in the constitution, which states that a minimum of 60% of Bhutan's total land should be maintained under forest cover for all time. The country even banned logging exports in 1999.

What's more, almost all the country's electricity comes from hydropower.

In fact, it produces so much hydroelectricity that it sells it to neighboring countries, which Bhutan claims offsets another 4.4 million tons of annual CO2 emissions. And Bhutan says that by 2025, increased hydroelectricity exports will let the country offset up to 22.4 million tons of CO2 per year in the region.

Source: CNN (https://edition.cnn.com/2018/10/11/asia/bhutan-carbon-negative/index.html)

The 2018 Environmental Performance Index (EPI) ranked Bhutan at 131 among other 180 countries, which accessed on 24 performance indicators across ten issue categories covering environmental health and ecosystem vitality (EPI, 2018). Green Environmental conservation is one of the four pillars of Bhutan's Gross National Happiness principles, and the government places high importance on sustainable and green initiatives. The potential impact on the economy is also significant: public procurement represents approximately 21 percent of GDP and 60–70 percent of the government's annual budget (IISD, 2017). The government of Bhutan has embraced the value of green public procurement (GPP) and is setting an example for other countries on how to leverage the power of the public purse to advance sustainable development.

The Green Public Procurement in Bhutan (GPP Bhutan) project was the EU-funded project under its EuropeAid SWITCH-Asia Program. The project was implemented by a consortium of two international and three national organizations, namely the International Institute for Sustainable Development (Project Lead, Canada), the Bhutan Chamber of Commerce and Industry (Bhutan), the Royal Institute of Management (Bhutan), the Royal Society for Protection of Nature (Bhutan) and the Collaborating Centre on Sustainable Consumption and Production (Germany).

The GPP Bhutan Project aimed to establish a strategic approach to scaling-up public demand for environmentally and socially preferable goods, services and infrastructure in Bhutan. The project sought to promote value for money across the asset lifecycle rather than merely at the point of purchase, in order to reduce the negative environmental impacts and maximize social benefits. Lama (2014) argues that sustainable or green public procurement (GPP) - whereby government authorities seek to procure goods, services and works with a reduced environmental impact throughout their life cycle, as well as with an increased positive social impact - has the potential to drive sustainable growth in Bhutan.

Moreover, Lama (2014) firmly believes that GPP is not only an instrument for sustainability and value for money but also an opportunity for economic development in Bhutan by promoting sustainable public works and infrastructure and developing local green goods and service industries.

Precursors for the GPP Bhutan were the following (but not limited to):

- Bhutan's development philosophy of Gross National Happiness;
- National Environment Commission's guiding strategy *The Middle Path*, which solicits a balanced approach to development;
- Bhutan's 11th Five-year Plan with the specific objective of "Self-reliance and inclusive green socio-economic development;
- The Economic Development Policy (EDP) – 2010 of Bhutan, which was geared towards green development; and

- The Cottage, Small and Medium Industry Policy (CSMI) 2012 having a vision to develop a CSMI sector in harmony with the GNH philosophy.

A country can achieve success regarding policy implementation only when the country already has got the foundation laid, and a spectrum of logically interlinked policies orient their goals to the same mission. While there were laws and regulations already in place to address and consider environmental and social aspects in the country, it has not been implemented in procurement. In the case of Bhutan, the policy framework for the green procurement worked pretty well as the leadership understood the value of the policies and strategies coordination for achieving the desired objectives. In this context, Green Public Procurement Bhutan Project was a strategic approach to look into the existing procurement legislation and enable it with green public procurement provisions to procure goods and services with a reduced environmental impact to bring in green governance in the country. In three and a half years, the government and the project partners helped to engage civil servants and local business and to integrate GPP into procurement practices.

UN (2018) revealed based on the Sustainable Consumption and Production (SCP) Bhutan report 2016, the average government procurement expenditure per year was about Nu. 18.2 billion, which accounts for 17.4 per cent of GDP in 2014. Given the high volume of procurement by the government, green elements are being incorporated into the public procurement system such as standard bidding documents with clear green component (pp. 60-61).

GPP BHUTAN PROJECT IMPLEMENTATION APPROACH

Not everywhere works the same implementation approach, but the approach taken by Bhutan from the start to end gives a big picture as well as nuances to be considered during the project implementation where the challenges are there from the concept on itself, and its track and behavior during implementation and outcome expectations are linked to humanity at large.

Activities taken under the GPP Bhutan project implementation approach are (but not limited to) (IISD, 2017):

1. Conduct the research necessary to start making public agencies, parliamentarians, policy-makers and suppliers aware of the potential of GPP.
2. Provide policy support for facilitating the introduction of GPP at all government levels in Bhutan.
3. Develop a range of materials and training to help public procurers to buy "green"

4. Select pilot projects with the City of Thimphu and the Ministry of Works and Human Settlements
5. Provide technical assistance on the financial viability of infrastructure projects and on the tender documents.

As part of the activities to make the designed approach successful, following complementary activities were carried out:

- Stakeholder engagement by conducting different levels of seminars targeting for parliamentarians, policymakers, public officials, public procurement entities, civil society, and private sector
- Develop policy recommendations on GPP in Bhutan
- Develop dedicated GPP guidance materials for public procurers with green criteria for common use items (e.g., air conditioners, apparel, paper) and criteria for green infrastructure (e.g., green roads, building and hydropower plants)
- Design preferential programs for small and medium enterprises and disadvantaged suppliers
- Design and facilitate GPP training sessions for public procurers and suppliers
- Mentor real-time GPP pilot tenders in selected industrial sectors
- Establish a GPP knowledge platform and curricula for long term knowledge sharing on GPP implementation

Leadership support at the policy level, as well as implementation level, always comes first for driving policy implementations. Policy implementations can be successful only if the bureaucrats support the initiative in action. Ministry of Finance of Bhutan was very proactive in support of the recommendations by endorsing officially in 2015. The initiative took a serious momentum when the parliament set up a working group to encourage GPP implementation in Bhutan.

CHECKLIST OF KEY CONSIDERATIONS FOR GPP IMPLEMENTATION

According to Lama & Casier (2017), implementing GPP does not happen overnight. Worldwide public procurement is gradually being deployed as a strategic government tool. Transitioning to GPP requires incremental changes to the public procurement laws, systems, and processes, but mostly to the mindset of the public procurers and policymakers. Before wrapping up the project, GPP Bhutan developed a comprehensive checklist[1] for implementing GPP in Bhutan as a guideline for public procurers and policymakers to get started on purchasing sustainably, as well as a gauge for Bhutan's progress in implementing GPP. This checklist can serve as an important guiding navigator while implementing GPP in any country. The checklist includes 19 questions, not necessarily to be followed in sequence. In fact, many of the questions can be reviewed in parallel except for few that are interdependent. Each of the items in the list are discussed briefly here. Understanding the purpose of these questions from the checklist provides confidence and necessary resources for the implementation of GPP.

1. Does the legal framework allow for green or sustainable public procurement?

Countries may not have mandated the green public procurement or sustainable procurement or even the provision of life-cycle consideration in the evaluation process for making procurement decisions. GPP implementation becomes more comfortable and faster if the procurement legislations are already amended or introduced through other legal instruments. Even if the procurement legislation does not have such provisions for the sustainable or green procurement, other laws issued for energy efficiency or recycled content, housing standards, environmental law, eco-label requirements, labor law, value for money clauses of the act and other provisions may also provide ground for buying green or sustainable products.

2. Are there soft laws and policies mandating green and sustainable procurement?

In many cases circular of the government, changes in the regulation or issuance of guidelines are also sufficient for procuring green products and apply green criteria.

3. To what extent are e-procurement methodologies being integrated into procurement and tender processes?

e-Procurement can automate end-to-end procurement processes to enhance efficiency, value for money, cost savings by eliminating the use of papers and standardization of administrative procedures, using innovative way of procurements like framework agreements, and helping design environment friendly products selection, setting green evaluation criteria, and tracking green performance of products and processes (Shakya, 2016). If the e-Procurement system is already in place, green procurement process becomes much easier to implement.

4. What do governments buy?

A register of items procured by the government on a frequent basis or in an ad hoc basis for different purposes with their annual volume and values should be prepared. Starting GPP criteria with the common-use items gives the result earlier. In some countries, even the GPP is launched from the large and complex PPP projects. For emerging countries, to minimize the risks of failure and give ample opportunity for the private sector be prepared for the product development in compliance with the green criteria, starting with common-use items like air conditioners, vehicles, cleaning agents, fuel, lightings, paper, furniture, IT equipment, uniform, etc. can be recommended.

5. Are green and sustainable products/services available in the domestic market?

The initial phase of GPP implementation is always challenging as the local market may not be ready to meet the requirement of the procuring entities. Hence, it is crucial to assess the local supply market for green products. Procuring entities should be promoting local innovations and market readiness for sustainable procurement.

6. How are budgets organized and managed?

According to Lama & Casier (2017), "how procurement budgets are organized has a lot to do with the extent to which sustainable public procurement will be a successful policy for sustainable development. Multi-year accounting and budget frameworks that allow temporal flexibility to carry over or borrow against the future are important to account for benefits that accrue during the useful life of assets. Such frameworks also include opportunities to integrate lifecycle costing and net present value into procurement decisions. Lifecycle costing and net present value are particularly important, as they will help procurement demonstrate that, although environmentally and socially preferable goods and services might cost more to purchase, they may be the most cost-efficient choice when looking at life cycle, bringing substantial savings during user life and end-of-life disposal. (If the environmental and social externalities are assigned monitory values and included in purchasing prices, then green goods and services become even more cost-effective)" (p. 13).

7. Allowing for lifecycle thinking and lifecycle costing in public procurement.

International Organization for Standardization standard, Buildings, and Constructed Assets, Service-Life Planning, Part 5: Lifecycle Costing (ISO 15686-5) defined the Lifecycle Costing as an economic assessment considering all agreed projected significant and relevant cost flows over a period of analysis expressed in monetary value. The projected costs are those needed to achieve defined levels of performance, including reliability, safety, and availability.

In GPP, the use of LCC facilitates the contract award process should not be solely based only on the financial basis but should consider all the financial and non-financial gains that are offered by environmentally and socially sustainable items.

Typical LCC analyses should consider:

- Purchasing costs and all associated costs, such as delivery, installation, commissioning and insurance;
- Operating expenses, including utility costs such as energy, water use, and maintenance costs;
- End-of-life costs, such as removal, recycling or refurbishment, and decommissioning;
- Longevity and warranty time frames of the asset.

Use of Lifecycle cost into the procurement process provides the procuring entities to achieve the best value for money across the life of the item and ensures the procurement of green and reliable products.

8. What to do when green or sustainable is more expensive?

There is a perception that the products procured through GPP criteria are expensive, which is right in many cases as the desired outcomes are more sustainable and contribute towards the better environment. Procuring entities and policymakers are reluctant to go green because of tight budgets, and ever-growing demands to deliver better services in critical areas such as education, health, water and sanitation, electricity, public transportation, and nutrition. In these circumstances, spending more on 'expensive, greener alternatives' can be perceived as an unnecessary luxury. In such situations, procurement methods like Framework agreement for bulk and frequent purchases, the e-reverse auction for cost reduction, outsourcing services, renting or leases can be used.

9. The issue of split responsibilities for capital and revenue budgets.

According to Lama & Casier (2017), "one of the difficulties in fully implementing green procurement is to reconcile capital and the revenue budget, which usually fall under the jurisdiction of different government departments or organizations. While procurement contracting might be the responsibility of one agency, budgets are controlled by another; and the use and maintenance of the product/service/development belong to yet another. As the benefits of GPP accrue during the project life and at its end disposal, those bearing the capital costs may not be the first to realize the benefits of sustainable alternatives" (p. 18).

10. The environmental and social dimensions of supplier qualification and appraisal.

In the case of green procurement, bidders are evaluated in compliance with the laws applicable for the environmental requirement, accreditations if required, ecolabel compliance, and declaration on environmental and social performance. Usually, suppliers are pre-qualified, and their products are verified on the compliance requirements.

11. Can tenders be evaluated on non-financial grounds?

At the matured stage of GPP, the weight of the non-financial evaluation criteria is higher than financial criteria to encourage value for money during the life of the product. Initially, the non-financial criteria should not be more than economic criteria to promote the gradual transition to green procurement.

12. Monitoring contracts after green tenders have been awarded.

After award green criteria should be stipulated in the contract to facilitate efficient monitoring of the contractor and product performance.

13. Determining where to start with GPP implementation.

It is recommended to start from the items more comfortable to comply with the green requirements at first and piggyback on the ongoing complementary initiatives on sustainable development.

14. Changing mindset from "best value for money" to "best value across the asset lifecycle."

The difference between the traditional procurement and green public procurement mainly on the value focus. In green public procurement, consideration of sustainable procurement and lifecycle cost come as the critical dimensions for the change of mindset of the procuring entities as well as the bidding community.

15. Embedding lifecycle thinking in procurement procedures.

Lama & Casier (2017) clarifies that "the changing mindsets requires procurers to appreciate the comparative environmental and social advantage of the products and services they are buying. Comparative advantage is determined through either product related or performance-related criteria. Product criteria stipulate the design or composition of a product, whereas performance criteria specify performance aspects such as thresholds of energy efficiency, increased recyclability or longer user life" (p. 24).

It further explains that in

Targeting outcomes during the user life of assets, performance-based criteria promote innovation and enable bidding companies to develop creative strategies to attain the level of performance being requested. Product-based criteria, on the other hand, do not spur sustainable design improvements to the same extent, though often they do involve a lesser degree of uncertainty and risk. Using life cycle analyses will help procurers and policymakers better appreciate the trade-off among these selection criteria and make the best decisions on a case-by-case basis. Thinking across the life cycle also becomes easier when the classic question, "What needs to be bought or what infrastructure needs to be built?" is replaced with, "What are the needs that need to be fulfilled?" This shift opens the door for developing performance-based specifications and purchasing integrated services, which are important incentives for sustainable design. (p. 24)

16. Do procurers have incentives to buy green?

Procurement executives engaged in the green procurement should be well rewarded and involved in different levels of policy interventions to motivate and appreciate their role and responsibilities. A comprehensive rewarding program based on their performance on green achievements should be a factor for administrative promotions and reward.

17. Provide for ongoing skills development.

GPP implementation is quite a learning process as the territory knowledge and experience is new to most of the countries. A sustainable knowledge and skill enhancement platform for all types of audiences, parliamentarians, policymakers, bureaucrats, procurement executives, bidding community, civil society, and relevant others, has to be developed for the sustainable success of GPP implementation.

18. Maintain an ongoing dialogue with listed suppliers.

Procuring entities should engage with the suppliers in a regular basis to communicate the GPP requirements, needed readiness on compliance, required innovations from the suppliers in the product, process, and performance for securing and executing contracts successfully.

19. Make provisions to purchase from small businesses.

Government and procuring entities should encourage local content, products, and skills, especially from the small business to promote sustainable procurement and development through local innovations, competitive development and making the market ready for the green and sustainable procurement the domestic and international markets.

GPP KNOWLEDGE PLATFORM

Bhutan not only laid the foundation of policies for the implementation of GPP but also developed a body of knowledge on GPP for the sustainable dissemination of acquired knowledge and skills to the broader public and targeted audiences.

The online Green Public Procurement Knowledge Platform (Figure 1) is a component of the GPP Bhutan Project is hosted under the Department of Finance and Business at Royal Institute of Management (RIM).

The platform publishes the vast amount of information and publications related to GPP Bhutan, practitioner and scholar relevant resources, and its activities. Easy availability of GPP related knowledge and information has helped the Bhutanese stakeholders, and also international visitors, learn details of implementation issues and approaches to resolve those issues.

The knowledge platform also provides a series of GPP Training Programs (Figure 2) with the training courses, short and long durations, starting with the basics to understand the Green Public Procurement to the advanced courses on Sustainable Supply Chain Management (SSCM) and tailor-made training programs.

Figure 1. Screenshot of GPP Bhutan Knowledge Platform
Source: http://gpp.rim.edu.bt

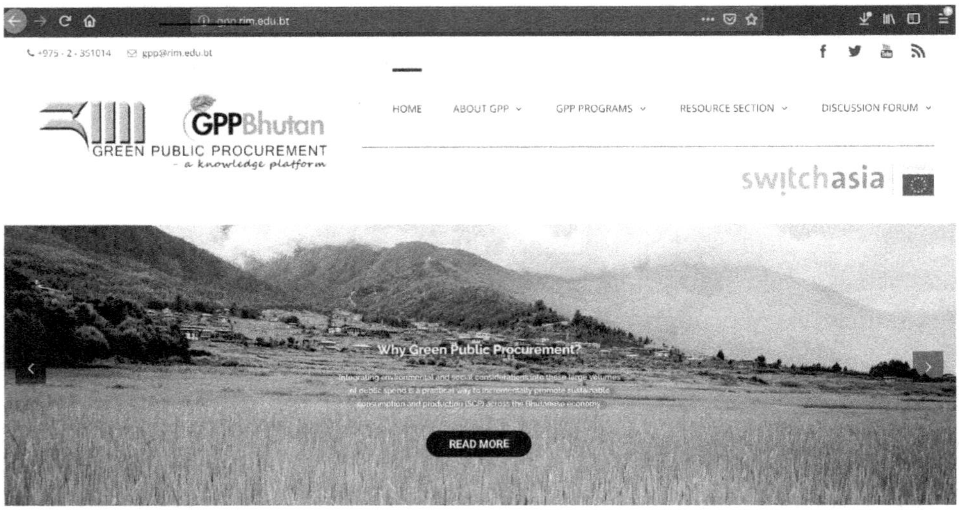

Figure 2. GPP Training Programs
Source: Source: http://gpp.rim.edu.bt

A Discussion Forum and GPP Blog form a platform for the citizen engagement facilitating expressing the voices of all stakeholders and posting and sharing of blogs on GPP relevant topics.

LESSONS LEARNED

GPP implementation needs the synergy of policy intervention by the leadership, mindset change regarding the value proposition of products, adequate engagement with procurers and suppliers, and knowledge dissemination. Ultimately, a shift towards GPP requires new processes, methods and criteria for procurement, re-training of procurers, availability of green alternatives, ability to assess the "greenness" of products and services using lifecycle costing and so on. For a developing nation like Bhutan, funding such a major overhaul in procurement is onerous, given the budget constraint and competing development needs of the country. Thus, if donors and development partners can be persuaded to include sustainability considerations to the deployment of their grants, this will serve as an essential incentive for GPP implementation in countries such as Bhutan (GPP Bhutan, 2015).

Assessing Bhutan's current stage of GPP against the checklist, it is clear that Bhutan's commitment to balanced and sustainable development through its Gross National Happiness philosophy provides a solid foundation for GPP. Bhutan's progress towards implementing GPP demonstrates early successes of positioning public procurement as a strategic tool to drive green industrial development, and ultimately switch to sustainable consumption and production patterns.

The following lessons can be learned from the GPP Bhutan project:

- Political will and leadership to support GPP vision and appropriate support from all stakeholders during implementation plays a critical role.
- GPP can be implemented even in absence of explicit legal provisions for green procurement. Complementary laws and regulations related to environment protection, sustainable development, energy standard, public health and so on can be used for building GPP policy foundation.
- Close and consistent engagement with key stakeholders (such as policymakers, public procurers, suppliers/producers and partners for GPP pilot project) is critical for achieving the change in mindset and gaining appreciation and acceptance of concepts such as lifecycle thinking, sustainable consumption and production, and environmental and social responsibility.
- Capacity building of both public procurers as well as suppliers is necessary. Both sides need a better understanding of: (1) what a "green" tender looks like, (2) how public procurers can engage with the suppliers in a transparent manner to be better informed about the market, and (3) how to respond to sustainability requirements in public tenders. (Casier, 2017)

- Cooperation between international and local partners is a great recipe for embedding the latest best practices on GPP into the local country context. Throughout the GPP Bhutan project, the ownership of knowledge on GPP and its implementation was transferred to the local partners, and embedded in the local institutions. It built the foundation for further knowledge sharing and capacity building in Bhutan for the years to come. This approach for implementing GPP is also replicable in other country contexts.
- Knowledge platform is an important resource bank that captures a complete spectrum of the body of knowledge, which can be a valuable asset for GPP implementation through the years.

REFERENCES

Casier, L. (2017). *Project Summary: How Green Public Procurement Became a Crucial Pillar for Bhutan's Sustainable Development Strategy*. Available at: https://www.iisd.org/sites/default/files/publications/gpp-bhutan-sustainable-development-strategy.pdf

EPI. (2018). *Environmental Performance Index Report. Yale Center for Environmental Law & Policy*. Yale University Center for International Earth Science Information Network, Columbia University. Available at: https://epi.envirocenter.yale.edu/downloads/epi2018policymakerssummaryv01.pdf

Forest, G. (2017). *Bhutan: Measuring Gross National Happiness over GDP*. Available at: https://www.theodysseyonline.com/bhutan-measuring-gross-national-happiness-gdp

GPP Bhutan. (2018). *Green Public Procurement Bhutan*. Available at http://gppbhutan.bt/

GPP Bhutan. (2015). *10 Recommendations for Moving Forward with Green Public Procurement in Bhutan*. Available at: http://gpp.rim.edu.bt/wp-content/uploads/2017/05/10-Recommendations_GPP-Bhutan.pdf

IISD. (2017). *Project Summary. How Green Public Procurement Became a Crucial Pillar for Bhutan's Sustainable Development Strategy*. Available at: https://www.iisd.org/sites/default/files/publications/gpp-bhutan-sustainable-development-strategy.pdf

Lama, P. (2014). *The case for Green Public Procurement in Bhutan: Achieving growth through sustainability*. Thimpu, Bhutan: Green Public Procurement in Bhutan (GPP Bhutan) Project. Available at http://www.gppbhutan.bt/sites/default/files/The_Case_fo_GPP_in_Bhutan_Web_Version.pdf

Lama, P., & Casier, L. (2017). *Checklist for Implementing Green Public Procurement in Bhutan*. Thimpu, Bhutan: The GPP Bhutan Project. Available at https://www.scp-centre.org/wp-content/uploads/2017/09/Checklist_for_Implementing_GPP_in_Bhutan.pdf

Shakya, R. K. (2016). *Digital Governance Digital Governance and E-Government Principles Applied to Public Procurement*. IGI Global.

Tutton, M., & Scott, K. (2018). *CNN: What tiny Bhutan can teach the world about being carbon negative*. Available at https://edition.cnn.com/2018/10/11/asia/bhutan-carbon-negative/index.html

UN. (2018). *Sustainable Development and Happiness.* Available at https://sustainabledevelopment.un.org/content/documents/19369Bhutan_NSDGR_Bhutan_2018.pdf

ENDNOTE

1 The checklist was adapted from an original framework developed by Oshani Perera at International Institute for Sustainable Development (IISD). Available at https://www.iisd.org/pdf/2011/procuring_green_public_sector.pdf

Compilation of References

AfDB. (2014). *Summary of Literature on Sustainable Public Procurement/Green Public Procurement.* African Development Bank. Available at https://www.afdb.org/fileadmin/uploads/afdb/Documents/Procurement/Project-related-Procurement/Summary_of_Literature_on_Sustainable_-_Green_Public_Procurement.pdf

Akenroye, T., Oyegoke, A., & Eyo, A. (2013). Development of a framework for the implementation of green public procurement in Nigeria. *International Journal of Procurement Management*, *6*(1), 1–23. doi:10.1504/IJPM.2013.050607

Alejandre, E., Traspaderne, V. A., & Ortiz de Elgea, V. A. (2010). Best practice on green or sustainable public procurement and new guidelines. Open House.

AMBA. (2016). *Lanzamiento de la Comisión Consultiva Del Área Metropolitana.* Retrieved July, 2017 from: https://www.gba.gob.ar/medios/noticias/lanzamiento_de_la_comisi%C3%B3n_consultiva_del_%C3%A1rea_metropolitana

Argentina General Environment Law 25.675. (2002). Retrieved July, 2017 from: http://ambiente.gob.ar/wp-content/uploads/POLITICA-AMBIENTAL-NACIONAL_Ley-25675.pdf

Argentina Renewable Energy Law 26.190 amend. by Law 27.191. (n.d.). Retrieved July, 2017 from: http://servicios.infoleg.gob.ar/infolegInternet/anexos/250000-254999/253626/norma.htm

Argentina. (2010). *National Census Portal.* Retrieved July, 2017 from: http://www.indec.gov.ar/definitivos_bajarArchivoNacionales.asp?idc=8&arch=x&c=2010

Argentina. (n.d.). *Argentina electrical emergency decree 134/15.* Retrieved July, 2017 from: https://www.boletinoficial.gob.ar/#!DetalleNormaBusquedaAvanzada/138328/20151217

Autonomous City of Buenos Aires. (2006). *Aires procurement and contracting law 2095/06*. Retrieved July, 2017 from: http://www.buenosaires.gob.ar/areas/leg_tecnica/sin/normapop09.php?id=91501

Autonomous City of Buenos Aires. (2011). *Autonomous City of Buenos Aires Mitigation and Adaption to Climate Change Law 3871/11*. Retrieved July, 2017 from: http://www.buenosaires.gob.ar/areas/leg_tecnica/sin/normapop09.php?id=181526

Basel. (n.d.). *Basel convention on the control of transboundary movements of hazardous wastes and their disposal*. Retrieved from: http://www.basel.int/

BCST. (2005). *National Research Council Board on Chemical Sciences and Technology, Sustainability in the Chemical Industry: Grand Challenges and Research Needs*. The National Academies Press. Available at http://www.nap.edu/catalog.php?record_id=11437

Becchetti, L., Bruni, L., & Zamagni, S. (2010). *Microeconomia*. Bologna: Il Mulino.

Bernroider, E. (2002). Factors in SWOT Analysis Applied to Micro, Small- to medium, and Large Software Enterprises: An Austrian Study. *European Management Journal*, *20*(5), 562–573. doi:10.1016/S0263-2373(02)00095-6

Bertoldi, P., Damian Bornás, C., Monni, S., & Piers de Raveschoot, R. (2010). *Linee guida "Come sviluppare un piano di azione per l'energia sostenibile – PAES"*. Unione Europea, Lussemburgo.

Bettini, P., Masini, D., & Ricotta, S. (2006) Acquistare, vendere e rispettare l'ambiente si può, in "Villaggio Globale", n° 36, Edizioni Villaggio Globale, Valenzano (Bari).

Bjorklund, M. (2011). Influence from the business environment on environmental purchasing: Drivers and hinders of purchasing green transportation services. *Journal of Purchasing and Supply Management*, *17*(1), 11–22. doi:10.1016/j.pursup.2010.04.002

Boromisa, A., Tišma, S. & Anastasya. (2016). *Green Jobs for Sustainable Development*. London, UK: Routledge.

Bouten, L., Everaert, P., Liedekerke, L. V., & De Moor, L. (2011). Corporate Social Responsibility Reporting: A Comprehensive Picture. *Accounting Forum*, *35*(3), 187–204. doi:10.1016/j.accfor.2011.06.007

Bouwer, M., de Jong, K., Jonk, M., Berman, T., Bersani, R., Lusser, H., . . . Szuppinger, P. (2005). *Green Public Procurement in Europe 2005 - Status overview. Virage Milieu & Management bv, Korte Spaarne 31, 2011 A.J Haarlem, the Netherlands.* Available at http://europa.eu.int/comm/environment/gpp/media.htm#state

Bowen, G. A. (2009). Document Analysis as a Qualitative Research Method. *Qualitative Research Journal*, *9*(2), 27–40.

Buenos Aires Ciudad. (n.d.). *Undersecretariat of operational management: Claudia Gabriela Jaime*. Retrieved from: http://www.buenosaires.gob.ar/hacienda/institucional-subsecretaria-de-gestion-operativa

Buenos Aires. (n.d.). *Climate change*. Retrieved from: http://www.buenosaires.gob.ar/areas/med_ambiente/apra/des_sust/pacc.php?menu_id=32408

Buenos Aires. (n.d.). Retrieved from: http://www.buenosaires.gob.ar/hacienda/compras/guia-para-el-consumo-y-compra-responsable-de-papel

C40. (n.d.). *C40 cities climates leadership group*. Retrieved from: http://www.c40.org/

Carbonn. (n.d.). Retrieved from: http://carbonn.org/

Carroll, A. B. (1979). A three-dimensional model of corporate social performance. *Academy of Management Review*, 4.

Casier, L. (2017). *Project Summary: How Green Public Procurement Became a Crucial Pillar for Bhutan's Sustainable Development Strategy*. Available at: https://www.iisd.org/sites/default/files/publications/gpp-bhutan-sustainable-development-strategy.pdf

Casier, L., Huizenga, R., Perera, O., Ruete, M., & Turley, L. (2015) *Implementing Sustainable Public Procurement in Latin America and the Caribbean. Handbook for the Inter-American Network on Government Procurement (INGP).* International Institute for Sustainable Development (IISD). Retrieved May, 2017 from: https://www.iisd.org/sites/default/files/publications/iisd-handbook-ingp-en.pdf

CDB. (n.d.). *Biological Diversity Agreement*. Retrieved from: https://www.cbd.int/

CDP. (n.d.). Retrieved from: https://data.cdp.net/

Cebeci, U. (2009). Fuzzy AHP-based decision support system for selecting ERP systems in textile industry by using balanced scorecard. *Expert Systems with Applications*, *36*(5), 8900–8909. doi:10.1016/j.eswa.2008.11.046

Chartered Institute of Procurement and Supply (CIPS). (2017). *Procurement Glossary website*. Accessed June, 2017 at: https://www.cips.org/knowledge/procurement-glossary/q/

Chen, J.-K., & Chen, I.-S. (2010). Using a novel conjunctive MCDM approach based on DEMATEL, fuzzy ANP and TOPSIS as an innovation support system for Taiwanese higher education. *Expert Systems with Applications*, *37*(3), 1981–1990. doi:10.1016/j.eswa.2009.06.079

Collage, D. (2009). *Carbon Neutral Action Report*. Available at https://www2.gov.bc.ca/assets/gov/environment/climate-change/cnar/2008/ps/douglas_college.pdf

Commission of the European Communities. (2008). *Public Procurement for a Better Environment*. Retrieved May, 2017 from: http://eurlex.europa.eu/LexUriServ/LexUriServ.do?uri=COM:2008:0400:FIN:EN:PDF

Commission of the European Communities. (2016). *Benefits of GPP*. Accessed July, 2017 at: http://ec.europa.eu/environment/gpp/benefits_en.htm

Constitution of the Argentine Nation §41 and §43. (n.d.). Retrieved July, 2017 from: http://www.biblioteca.jus.gov.ar/argentina-constitution.pdf

Constitution of the City if Buenos Aires §8. (n.d.). Retrieved July, 2017 from: http://www.buenosaires.gob.ar/areas/leg_tecnica/sin/normapop09.php?id=26766&qu=c

Court of Justice of the European Union (CJEU). (2002). Case No. C-513/99. Concordia Bus Finland v Helsingin Kaupunki (Concordia Bus). ECR I-7123.

Court of Justice of the European Union (CJEU). (2003). Case No. C-448/01. EVN AG and Wienstrom GmbH v Austria (Wienstrom). ECR I-4527.

Defra. (2008). *A framework for pro-environmental behaviors*. London: Department for Environment, Food and Rural Affairs Report. Available at https://www.gov.uk/government/uploads/system/uploads/attachment_data/file/69277/pb13574-behaviours-report-080110.pdf

Department of Expenditure and Reform. (2014). *Green Tenders: An Action Plan on Green Public Procurement*. Available at http://etenders.gov.ie/Media/Default/SiteContent/LegislationGuides/13.%20Green%20Tenders%20-%20An%20Action%20Plan%20on%20Green%20Public%20Procurement.pdf

Dinçer, Ö. (2004). *Stratejik Yönetim ve İşletme Politikası. 7*. İstanbul: Baskı, Beta Yayınları.

Dyson, R. G. (2004). Strategic development and SWOT analysis at the University of Warwick. *European Journal of Operational Research*, *152*(3), 631–640. doi:10.1016/S0377-2217(03)00062-6

EC. (2008). *A toolkit developed for the European Commission Local Governments for Sustainability. ICLEI.* Available at: http://ec.europa.eu/environment/gpp/pdf/toolkit/module2_legal_framework.pdf

Ecolex. (n.d.). *Framework agreement on the environment of MERCOSUR*. Retrieved from: http://www2.ecolex.org/server2neu.php/libcat/docs/TRE/Full/En/TRE-153663.pdf

Elkington, J. (1997). *Cannibal with forks: the Triple bottom line of 21st Century Business*. Oxford, UK: Capstone.

EPI. (2018). *Environmental Performance Index Report.* Yale Center for Environmental Law & Policy, Yale University Center for International Earth Science Information Network, Columbia University. Available at: https://epi.envirocenter.yale.edu/downloads/epi2018policymakerssummaryv01.pdf

EPI. (2018). *Environmental Performance Index Report. Yale Center for Environmental Law & Policy*. Yale University Center for International Earth Science Information Network, Columbia University. Available at: https://epi.envirocenter.yale.edu/downloads/epi2018policymakerssummaryv01.pdf

EU Commission. (2016). *Commission of the European Communities. Eco-labels*. Accessed July, 2017 at: http://ec.europa.eu/environment/gpp/eco_labels.htm

European Commission (EC). (2016). *Buying Green! - A Handbook on Green Public Procurement*. EC.

European Commission. (1996). Green Paper - Public procurement in the European Union: Exploring the Way Forward. EC.

European Commission. (2001a). Green Paper - Promoting a European framework for Corporate Social Responsibility, COM(2001) 366. EC.

European Commission. (2001b). Communication from the Commission to the Council, the European Parliament, the Economic and Social Committee and the Committee of the Regions, "Environment 2010: Our future, our choice, 6th Environmental Action Programme", COM(2001) 31. EC.

European Commission. (2001c). Communication from the Commission to the Council, the European Parliament, the Economic and Social Committee and the Committee of the Regions, "Interpretative communication of the Commission on the Community law applicable to public procurement and the possibilities for integrating environmental considerations into public procurement", COM(2001) 274. EC.

European Commission. (2003). Communication from the Commission to the Council and the European Parliament, "Integrated Product Policy Building on Environmental Life-Cycle Thinking", COM(2003) 302. EC.

European Commission. (2004a). Directive 2004/17/EC of the European Parliament and of the Council of 31 March 2004, "coordinating the procurement procedures of entities operating in the water, energy, transport and postal services sector" (2004/17/EC). *Official Journal of the European Union*, 1-113.

European Commission. (2004b). Directive 2004/18/EC of the European Parliament and of the Council of 31 March 2004, "coordinating the procedures for the award of public works contracts, public supply contracts and public service contracts" (2004/18/EC). *Official Journal of the European Union*, 114-240.

European Commission. (2008a). Communication from the Commission to the Council, the European Parliament, the Economic and Social Committee and the Committee of the Regions, "on the Sustainable Consumption and Production and Sustainable Industrial Policy Action Plan", COM(2008) 397. EC.

European Commission. (2008b). Communication from the Commission to the Council, the European Parliament, the Economic and Social Committee and the Committee of the Regions, "Public procurement for a better environment", COM(2008) 400. EC.

European Commission. (2009). Communication from the Commission to the Council, the European Parliament, the Economic and Social Committee and the Committee of the Regions, "Mainstreaming sustainable development into EU policies: 2009 Review of the European Union Strategy for Sustainable Development", COM(2009) 400. EC.

European Commission. (2010). Communication from the Commission, "Europe 2020. A strategy for smart, sustainable and inclusive growth", COM (2010) 2020 final, 3 March 2010. EC.

European Commission. (2011a). Direttiva del Parlamento Europeo e del Consiglio on Public Procurement, COM(2011) 896. EC.

European Commission. (2011b). *Commission staff working paper, Evaluation Report, "Impact and Effectiveness of EU Public Procurement Legislation", SEC(2011) 853*. EC.

European Commission. (2011c). Communication from the Commission to the Council, the European Parliament, the Economic and Social Committee and the Committee of the Regions, "Energy Roadmap 2050", COM(2011) 885. EC.

European Commission. (2011d). Buying Green! A Handbook on Green Public Procurement. EC.

European Commission. (2012). *Green Public Procurement, A collection of good practices*. EC.

European Commission. (2014a). Retrieved from http://ec.europa.eu/environment/gpp

European Commission. (2014b). Retrieved from http://www.europarl.europa.eu

European Commission. (2014c). Directive 2014/23/EU of the European Parliament and of the Council of 26 February 2014 on the award of concession contracts. *Official Journal of the European Union (28/03/2014)*, 1-64.

European Commission. (2014d). Directive 2014/24/EU of the European Parliament and of the Council of 26 February 2014 on public procurement and repealing Directive 2004/18/EC. *Official Journal of the European Union (28/03/2014),* 1-178.

European Commission. (2014e). Directive 2014/25/EU of the European Parliament and of the Council of 26 February 2014 on procurement by entities operating in the water, energy, transport and postal services sectors and repealing Directive 2004/17/EC. *Official Journal of the European Union (28/03/2014)*, 1-132.

European Commission. (2016). *EU GPP Criteria. Environment*. Author.

Expert Choice. (2000). *Expert Choice, Analytical Hierarchy Process (AHP) Software. Version 9.5*. Expert Choice.

Fiorani, G., Jannelli, R., & Meneguzzo, M. (2012). *CSR 2.0 Proattiva e sostenibile*. Milano: Egea.

Forest, G. (2017). *Bhutan: Measuring Gross National Happiness over GDP*. Available at: https://www.theodysseyonline.com/bhutan-measuring-gross-national-happiness-gdp

Garuti, C., & Spencer, I. (2007). Parallels between the analytic hierarchy and network processes (AHP/ANP) and fractal geometry. *Mathematical and Computer Modelling*, *46*(7-8), 926–934. doi:10.1016/j.mcm.2007.03.029

General Services Administration (GSA). (2007). *Sustainable Facilities Tool*. Accessed July, 2017 at: https://sftool.gov/learn/about/41/sustainability

GLCNSP. (n.d.). *Global Lead City Network on Sustainable Procurement*. Retrieved from: http://www.glcn-on-sp.org/home/

Global Eco-labeling Network. (2012). *Environmental labels and declarations: How ISO standards help*. International Organization for Standardization (ISO). Retrieved May, 2017 from: https://www.iso.org/files/live/sites/isoorg/files/archive/pdf/en/environmental-labelling.pdf

Glosbe. (n.d.). *Environmentally sound behavior*. Available at https://glosbe.com/en/fr/environmentally%20sound%20behaviour

Government of the City of Buenos Aires. (2015). *Climate Change Action Plan 2016-2020*. Retrieved July, 2017 from: http://cdn2.buenosaires.gob.ar/espaciopublico/apra/pacc_2020.pdf

Government of the City of Buenos Aires. (2017). *City of Buenos Aires Statistical System (SEC) (statistics portal)*. Accessed July, 2017 at: http://www.estadisticaciudad.gob.ar/eyc/

Government of the Commonwealth of Australia. (2013). *Department of Sustainability, Environment, Water, Population and Communities. Sustainable Procurement Guide*. Government of the Commonwealth of Australia. Retrieved June, 2017 from: https://www.environment.gov.au/system/files/resources/7b8df2bd-3bb9-49cc-b417-5f2eb6e0ce37/files/sustainable-procurement-guide.pdf

GPP Bhutan. (2015). *10 Recommendations for Moving Forward with Green Public Procurement in Bhutan*. Available at: http://gpp.rim.edu.bt/wp-content/uploads/2017/05/10-Recommendations_GPP-Bhutan.pdf

GPP Bhutan. (2018). *Green Procurement in Bhutan.* Available at http://gppbhutan.bt/

GPP Bhutan. (2018). *Green Public Procurement Bhutan*. Available at http://gppbhutan.bt/

Green Audit. (2017). *What is Green Audit?* Available at: https://www.greenaudit.org/about/

Green Public Procurement and the European Eco label, Fact sheet. (2008). Toolkit developed for the European Commission by ICLEI - Local Governments for Sustainability, European Commission, DG Environment-G2, B-1049.

Green Public Procurement. (2012). *A collection of good practices*. Luxembourg: Publications Office of the European Union. Retrieved from http://europa.eu/environment/gpp/)

Green, B. (2016). *A hand book on public procurement* (3rd ed.). Luxembourg: Publications Office of the European Union. doi:10.2779/246106

Grinier, S. (2014). *The INGP and the Latin-American Outlook on Sustainable Public Procurement.* INGP Annual Meeting. Retrieved May, 2017 from: http://ricg.org/download.php?len=es&id=293&nbre=1%20Presentation%20CPS%20RICG-2014%20%20Steven%20Griner%20Brasilia%20-%20Version%20Corta1.ppt&ti=application/vnd.ms-powerpoint&tc=Contenidos

Harrison, J. P. (2010). *Strategic planning and Swot Analysis, Essentials of Strategic Planning in Healthcare.* Health Administration Press. Available at http://www.ache.org/pdf/secure/gifts/harrison_chapter5.pdf

HSE. (2018). *Implementation of Green Procurement for HSE Procurement/Supplies Officers.* Available at https://www.hse.ie/eng/services/news/newsfeatures/healthsustainabilityoffice/greenprocurement/gpp%20factsheet.pdf

IISD. (2017). *Project Summary. How Green Public Procurement Became a Crucial Pillar for Bhutan's Sustainable Development Strategy.* Available at: https://www.iisd.org/sites/default/files/publications/gpp-bhutan-sustainable-development-strategy.pdf

Indiana University. (2018). *Sustainable Purchasing*. Available at http://www.indiana.edu/~purchase/sustainability/sustainability.php

Indira C. M. (2016). Green banking practise in India. *International Journal of Advanced Research, 4*(11). doi:10.21474/IJAR01/2156

International Institute for Sustainable Development (IISD). (2013). *Business and Sustainable Development: A Global Guide*. Retrieved June, 2017 from: https://www.iisd.org/business/tools/bt_green_pro.aspx

Jackson, P. (2012). *Value for money and international development: Deconstructing myths to promote a more constructive discussion*. OECD. Retrieved June, 2017 from: http://www.oecd.org/development/effectiveness/49652541.pdf

Kahraman, C., Demirel, N. Ç., & Demirel, T. (2007). Prioritization of e-Government strategies using a SWOT-AHP analysis: The case of Turkey. *European Journal of Information Systems*, *16*(3), 284–298. doi:10.1057/palgrave.ejis.3000679

Kajanus, M., Kangas, J., & Kurttıla, M. (2004). The use of value focused thinking and the A SWOT hybrid method in tourism management. *Tourism Management*, *25*(4), 499–506. doi:10.1016/S0261-5177(03)00120-1

Kurtilla, M., Pesonen, M., Kangas, J., & Kajanus, M. (2000). Utilizing the analytic hierarchy process (AHP) in SWOT analysis - a hybrid method and its application to a forest certification case. *Forest Policy and Economics*, *1*(1), 41–52. doi:10.1016/S1389-9341(99)00004-0

Lama, P. (2014). *The case for Green Public Procurement in Bhutan: Achieving growth through sustainability*. Thimpu, Bhutan: Green Public Procurement in Bhutan (GPP Bhutan) Project. Available at http://www.gppbhutan.bt/sites/default/files/The_Case_fo_GPP_in_Bhutan_Web_Version.pdf

Lama, P., & Casier, L. (2017). *Checklist for Implementing Green Public Procurement in Bhutan*. Thimpu, Bhutan: The GPP Bhutan Project. Available at https://www.scp-centre.org/wp-content/uploads/2017/09/Checklist_for_Implementing_GPP_in_Bhutan.pdf

Lee, A. H. I., Chen, W.-C., & Chang, C.-J. (2008). A fuzzy AHP and BSC approach for evaluating performance of IT department in the manufacturing industry in Taiwan. *Expert Systems with Applications*, *34*(1), 96–107. doi:10.1016/j.eswa.2006.08.022

Lee, S. F., & Lo, K. K. (2003). e-Enterprise and management course development using strategy formulations framework for vocational education. *Journal of Materials Processing Technology*, *139*(1-3), 604–612. doi:10.1016/S0924-0136(03)00501-6

Leskınen, L. A., Leskınen, P., Kurttıla, M., Kangas, J., & Kajanus, M. (2006). Adapting modern strategic decision support tools in the participatory strategy process-a case study of a forest research station. *Forest Policy and Economics*, *8*(3), 267–278. doi:10.1016/j.forpol.2004.06.007

Made-by. (2018). *EU Ecolabel for Textiles. Made-By.org*. Available at: http://www.made-by.org/consultancy/standards/eu-ecolabel-for-textiles/

Marras. (2008). *The United Nations Sustainable Procurement Guide*. United Nations Environment Programme (UNEP). Retrieved July, 2017 from: http://www.greeningtheblue.org/sites/default/files/Buyingforabetterworld_0.pdf

Marron, D. (2014). Greener Public Purchasing as an Environmental Policy Instrument. *OECD Journal on Budgeting*, 3(4), 71-102. doi:10.1787/budget-v3-art23-en

Masozera, M. K., Alavalapatı, J. R. R., Jacobson, S. K., & Shresta, R. K. (2004). *Assessing the suitability of community-based management for the Nyungwe Forest Reserve*. Forest Policy and Economics.

MATTM. (2013). Piano d'Azione per la sostenibilità ambientale dei consumi nel settore della Pubblica Amministrazione – Revisione 2013. *Gazzetta Ufficiale della Repubblica Italiana*, *102*, 17–29.

Maxwell, D. (2008). Sustainable Clothing Action Plan, Department for Environment. *Lead Sustainable Clothing Roadmap, Food and Rural Affairs*, 2-11. Available at http://glasaaward.org/wp-content/uploads/2014/01/Clothing-Actionplan.pdf

Mendes, R. (2017). *Health in the green economy: Co-benefits to health of climate change mitigation.* World Health organization (WHO). Retrieved May, 2017 from: http://www.who.int/hia/green_economy/hgebrief_occ.pdf

Meneguzzo. (2012). *Managerialità innovazione e governance*. Archidata.

Ministry of Energy. (n.d.). *Renewable energy law*. Retrieved from: http://www.energia.gob.ar/contenidos/verpagina.php?idpagina=3876

NAPP. (2015). *Benefit from public procurement*. Available at: http://www.upphandlingsmyndigheten.se/en/sustainable-public-procurement/why-sustainable-procurement/beneficial-for-society/

NEMA. (2018). *National Environment Management Authority*. Available at https://www.nema.go.ke/index.php?option=com_content&view=article&id=1&Itemid=136

NEMC. (2018). *National Environment Management Council*. Available at https://www.unenvironment.org/nairobiconvention/national-environment-management-council-nemc-tanzania

OECD. (1996). *Ministerial Symposium on the future of public services*. OECD/PUMA.

OECD. (2001). *OECD Environmental Strategy for the First Decade of the 21st Century*. Retrieved from http://www.oecd.org/environment/environmentalindicatorsmodellingandoutlooks/1863539.pdf

OECD. (2009). *Cities, Climate Change and Multilevel Governance*. OECD Environmental Working Papers n°14, 2009. OECD Publishing.

OECD. (2013). *Government at a Glance 2013*. OECD Publishing.

OECD. (2015). *Going green: Best practices for sustainable procurement*. Secretary-General of the OECD. Retrieved May, 2017 from: http://www.oecd.org/gov/ethics/Going_Green_Best_Practices_for_Sustainable_Procurement.pdf

ONU. (2013). Retrieved from http://unstats.un.org/unsd/methods/m49/m49regin.htm#europe

Organisation for Economic Co-operation and Development (OECD). (2015). *OECD Recommendation of the Council on Public Procurement*. OECD.

Pan, N.-F. (2008). Fuzzy AHP approach for selecting the suitable bridge construction method. *Automation in Construction*, *17*(8), 958–965. doi:10.1016/j.autcon.2008.03.005

Pearce, D., Markandya, A., & Barbier, E. B. (1989). *Blueprint for a Green Economy*. London: Earthscan.

Perera, O. (2011). *Procuring Green in the Public Sector: A checklist for getting started*. Available at: https://www.iisd.org/pdf/2011/procuring_green_public_sector.pdf

Perera, O. (2015). *Implementing Sustainable Public Procurement in Latin America and the Caribbean. Handbook for the Inter-American Network on Government Procurement (INGP)*. International Institute for Sustainable Development (IISD),. Retrieved May, 2017 from: https://www.iisd.org/pdf/2009/life_cycle_costing.pdf

PPF. (2017). *Report on the 10th East African Procurement Forum*. PPADB, Kampala, Uganda.

PPRC-Kosovo. (2014). *National Public Procurement Strategy.* Available at https://krpp.rks-gov.net/krpp/PageFiles/File/draftLS/2014/00-National%20Public%20Procurement%20Strategy-v3_ENG.docx

PricewaterhouseCoopers. (2009). *Collection of statistical information on Green Public Procurement in the EU*. Author.

Promentilla, M. A. B., Furuichi, T., Ishii, K., & Tanikawa, N. (2008). A fuzzy analytic network process for multi-criteria evaluation of contaminated site remedial countermeasures. *Journal of Environmental Management*, *88*(3), 479–495. doi:10.1016/j.jenvman.2007.03.013 PMID:17467879

Public Services and Procurement Canada (PSPC). (2006). *Policy on Green Procurement.* Government of Canada. Retrieved May, 2017 from: http://www.tpsgc-pwgsc.gc.ca/ecologisation-greening/achats-procurement/politique-policy-eng.html

Public-Private Infrastructure Advisory Facility (PPIAF). (2013). *Value-for-Money Analysis Practices and Challenges: How Governments Choose When to Use PPP to Deliver Public Infrastructure and Services* (Report from World Bank Global Round-Table). The World Bank Institute. Retrieved May, 2017 from: https://openknowledge.worldbank.org/bitstream/handle/10986/17622/840800WP0Box380ey0Analysis00PUBLIC0.pdf;sequence=1

PwC. (2009). *Collection of Statistical Information on Green Public Procurement in the EU.* Available at http://ec.europa.eu/environment/gpp/pdf/statistical_information.pdf

Qiao, Y., & Wang, C. (2011). Issues and Challenges in Implementing China's Green Public Procurement Program. *Journal of Environmental Protection*, *2*(08), 1034–1045. doi:10.4236/jep.2011.28119

Ramsar. (n.d.). *Ramsar convention on wetlands of international importance*. Retrieved from: http://www.ramsar.org/about-the-ramsar-convention

Ricotta, S. (2012). Il Green Public Procurement per il cambiamento climatico. Gazzetta Ambiente, 2.

Roose, R., & de la Harpe, S. (2008). Good governance in public procurement: a South African case study. *PER: Potchefstroomse Elektroniese Regsblad, 11*(2). Available at http://www.scielo.org.za/scielo.php7pidM1727-37812008000200006&script-sciarttext

Saaty, T. L., & Sodenkamp, M. (2008). Making decisions in hierarchic and network systems. *International Journal Applied Decision Sciences, 1*(1), 24-56. available at http://www.bitstoenergy.ch/wpcontent/uploads/2016/01/2008_IJADS_Saaty_Sodenkamp.pdf

Saaty, T. L. (1980). *The Analytic Hierarchy Process*. McGraw-Hill International Book Company.

Saaty, T. L. (1991). Some Mathematical Concepts of the Analytic Hierarchy Process. *Behaviormetrika*, *29*(29), 1–9. doi:10.2333/bhmk.18.29_1

Saaty, T. L. (2007). Time dependent decision-making; dynamic priorities in the AHP/ANP: Generalizing from points to functions and from real to complex variables. *Mathematical and Computer Modelling*, *46*(7-8), 860–891. doi:10.1016/j.mcm.2007.03.028

Saltari, E., & Travaglini, G. (2005). *Le radici del declino economico*. Utet Università.

Secretariat of the Antarctic Treaty. (n.d.). *Protocol on Environmental Protection to the Antarctic*. Retrieved from: http://www.ats.aq/

Seker, S., & Özgürler, M. (2012). Analysis of the Turkish consumer electronics firm using SWOT-AHP method. *Procedia: Social and Behavioral Sciences*, *58*, 1544–1554. doi:10.1016/j.sbspro.2012.09.1141

SELA. (2015). *Public procurement as a tool for development in Latin America and the Caribbean*. Permanent Secretariat of SELA. Retrieved May, 2017 from: http://www.sela.org/media/268509/public-procurement-as-a-tool-for-development-in-lac.pdf

Shakya, R. K. (2016). *Digital Governance Digital Governance and E-Government Principles Applied to Public Procurement*. IGI Global.

Shakya, R. K. (2016). *Green Public Procurement Strategy for East African Community (EAC)*. Arusha, Tanzania: EAC.

Shakya, R. K. (2016). *Green Public Procurement Strategy for the East African Community (EAC)*. Arusha, Tanzania: EAC.

Shannuganthan, P., & Warren, L. M. (1997). Status of Sustainable Development as a Norm of International Law. *Journal of Environmental Law*, *9*, 221.

Shaver, J. A. (1999). Applying ISO 14001 to environmental management. *Textile Chemist and Colorist*, 27–30.

Shrestha, R. K., Alavalapati, J. R. R., & Kalmbacher, R. S. (2004). Exploring the potential for silvopasture adoption in south-central Florida: An application of SWOT–AHP method. *Agricultural Systems*, *81*(3), 185–199. doi:10.1016/j.agsy.2003.09.004

Sieb, A. (2011). *"Five benefits of green procurement" E-sourcing forum.* Determine Inc. Retrieved May, 2017 from: http://www.esourcingforum.com/archives/2011/11/29/five-benefits-of-green-procurement/

Siemens, R. (2003). A Review and critical evaluation of selected greener public purchasing programmes and policies. In *The Environmental Performance of Public Procurement. Issues of Policy Coherence* (pp. 51–95). Available at http://www.keepeek.com/Digital-Asset-Management/oecd/environment/the-environmental-performance-of-public-procurement_9789264101562-en#page50

Šličiuvienė, D. (2017). Public procurement in Lithuania: (dis)balance between profitability and environmental protection. In K. V. Thai (Ed.), *Global Public Procurement Theories and Practices, Public Administration, Governance and Globalization* (pp. 17–38). New York, NY: Springer. doi:10.1007/978-3-319-49280-3_2

Stewart, R., Moarmed, S., & Daet, R. (2002). Strategic implementation of IT/IS projects in construction: A case study. *Automation in Construction*, *11*(6), 681–694. doi:10.1016/S0926-5805(02)00009-2

Strandberg, C., & Robinson, A. (2008). *Business Case & Benefits of Sustainability Purchasing*. The BuySmart Network. Retrieved May, 2017 from: http://www.buysmartbc.com/_Library/Resources/resource_bsn_business_case_to_sustainability_2008.pdf

Testa, F., Iraldo, F., Frey, M., & Daddi, T. (2012). What factors influence the uptake of GPP (green public procurement) practices? New evidence from an Italian survey. *Econological Economics, 82*(2012), 88-96. Available at http://ec.europa.eu/environment/gpp/pdf/uptake_GPP_practices.pdf

The Public Contracts Regulations. (2015). Available at https://link.springer.com/chapter/10.1007/978-3-319-49280-3_2/fulltext.html

The World Bank. (2013). Retrieved from: https://openknowledge.worldbank.org/bitstream/handle/10986/17622/840800WP0Box380ey0Analysis00PUBLIC0.pdf;sequence=1

Tukker, Emmert, S., Charter, M., Vezzoli, C., Sto, E., Munch Andersen, M., ... Lahlou, S. (2008). Fostering change to sustainable consumption and production: An evidence based view. *Journal of Cleaner Production*, *16*(11), 1218–1225. doi:10.1016/j.jclepro.2007.08.015

Turkey's Sustainable Development Report: Claiming the Future Best Practice Examples on Sustainable Development. (n.d.). Available at http://www.undp.org/content/dam/turkey/docs/Publications/EnvSust/2.Best%20Practice%20Examples.pdf

Tutton, M., & Scott, K. (2018). *CNN: What tiny Bhutan can teach the world about being carbon negative*. Available at https://edition.cnn.com/2018/10/11/asia/bhutan-carbon-negative/index.html

UCCI. (n.d.). *Lo nuevo de la ucci*. Retrieved from: http://ciudadesiberoamericanas.org/

Ülgen, H., & Mirze, S. K. (2004). *İşletmelerde Stratejik Yönetim*. İstanbul: Literatür Yayınları.

UN Climate Change. (n.d.). *Paris Climate Accord*. Retrieved from: http://newsroom.unfccc.int/paris-agreement/

UN Environment DTIE. (2012). *The Impacts of Sustainable Procurement: Eight illustrative case studies*. Retrieved from http://www.unep.fr/scp/procurement/docsres/projectinfo/studyonimpactsofspp.pdf

UN. (2018). *Sustainable Development and Happiness*. Available at https://sustainabledevelopment.un.org/content/documents/19369Bhutan_NSDGR_Bhutan_2018.pdf

UNCCD. (n.d.). *United Nations convention to combat desertification*. Retrieved from: http://www2.unccd.int/

UNCED. (1992). Agenda 21. *United Nations Conference on Environment & Development*, Rio de Janeiro, Brazil.

UNDESA. (2002). *Global Challenge Global Opportunity: Trends in sustainable Development*. Available at https://sustainabledevelopment.un.org/content/documents/critical_trends_report_2002.pdf

UNEP. (2001). *Montreal Protocol on Substances that Deplete the Ozone Layer*. Retrieved from: http://ozone.unep.org/en/treaties-and-decisions/montreal-protocol-substances-deplete-ozone-layer

UNEP. (2011). *Buying for the better world: A guide on sustainable procurement for the UN system*. Paris, France: UNEP.

UNEP. (n.d.). *Vienna convention for the protection of the ozone layer*. Retrieved from: http://ozone.unep.org/en/treaties-and-decisions/vienna-convention-protection-ozone-layer

UNESCAPE. (2012). *Low Carbon Green Growth Roadmap for Asia and the Pacific*. Available at http://www.unescap.org/sites/default/files/33.%20FS-Green-Public-Procurement.pdf

UNFCCC. (n.d.). *Kyoto protocol – targets for the first commitment period*. Retrieved from: http://unfccc.int/kyoto_protocol/items/2830.php

United Nations General Assembly. (2015). *Transforming Our World: the 2030 Agenda for Sustainable Development.* A/RES/70/1. UN.

United Nations. (2017). *The sustainable development agenda 2030*. Retrieved from: http://www.un.org/sustainabledevelopment/development-agenda/

Vahidnia, M. H., Alesheikh, A. A., & Alimohammadi, A. (2009). Hospital site selection using fuzzy AHP and its derivatives. *Journal of Environmental Management*, *90*(10), 3048–3056. doi:10.1016/j.jenvman.2009.04.010 PMID:19477577

WHC. (1972). *Convention concerning the protection of the world cultural and natural heritage*. Retrieved from: http://whc.unesco.org/en/conventiontext/

Wikipedia. (2017). *List of Metropolitan Areas in the Americas.* Accessed July, 2017 at: https://en.wikipedia.org/wiki/List_of_metropolitan_areas_in_the_Americas

Williams-Elegbe, S. (2017). *Public Procurement and Multilateral Development Banks: Law, Practice and Problems*. Portland, OR: Hart Publishing.

World Commission on Environment and Development. (1987). *Our Common Future*. Oxford University Press.

World Trade Organization (WTO). (2014). *General Agreement on Public Procurement (GPA)*. WTO.

Yaker, F. (2013). *Sustainable Public Procurement: A Global Review Final Report.* United Nations Environment Programme (UNEP). Retrieved May, 2017 from: https://www.globalecolabelling.net/assets/Documents/unep-spp-report.pdf

Yang, C.-L., Chuang, S.-P., & Huang, R.-H. (2009). Manufacturing evaluation system based on AHP/ANP approach for wafer fabricating industry. *Expert Systems with Applications*, *36*(8), 11369–11377. doi:10.1016/j.eswa.2009.03.023

Yüksel, İ., & Dağdeviren, M. (2007). Using the analytic network process (ANP) in a SWOT analysis – A case study for a textile firm. *Information Sciences*, *177*(16), 3364–3382. doi:10.1016/j.ins.2007.01.001

Yüksel, İ., & Dağdeviren, M. (2010). Using the fuzzy analytic network process (ANP) for Balanced Scorecard (BSC): A case study for a manufacturing firm. *Expert Systems with Applications*, *37*(2), 1270–1278. doi:10.1016/j.eswa.2009.06.002

Zuppinger, P. (2005). *Green Public procurement in Europe 2005 - Status Overview.* Haarlem: Virage Milieu & Management.

Related References

To continue our tradition of advancing academic research, we have compiled a list of recommended IGI Global readings. These references will provide additional information and guidance to further enrich your knowledge and assist you with your own research and future publications.

Abresch, J. (2003). Geographic Information Systems Research and Data Centers. In A. Hanson & B. Levin (Eds.), *Building a Virtual Library* (pp. 52–64). Hershey, PA: IGI Global. doi:10.4018/978-1-59140-106-3.ch004

Ajiboye, O. E., & Yusuff, O. S. (2017). Foreign Land Acquisition: Food Security and Food Chains – The Nigerian Experience. In I. Management Association (Ed.), Natural Resources Management: Concepts, Methodologies, Tools, and Applications (pp. 1524-1545). Hershey, PA: IGI Global. doi:10.4018/978-1-5225-0803-8.ch072

Alapiki, H. E., & Amadi, L. A. (2018). Sustainable Food Consumption in the Neoliberal Order: Challenges and Policy Implications. In A. Obayelu (Ed.), *Food Systems Sustainability and Environmental Policies in Modern Economies* (pp. 90–123). Hershey, PA: IGI Global. doi:10.4018/978-1-5225-3631-4.ch005

Altaş, A. (2018). Geographical Information System Applications Utilized in Museums in Turkey Within the Scope of the Cultural Heritage Tourism: A Case Study of Mobile Application of Müze Asist. In S. Chaudhuri & N. Ray (Eds.), *GIS Applications in the Tourism and Hospitality Industry* (pp. 42–60). Hershey, PA: IGI Global. doi:10.4018/978-1-5225-5088-4.ch002

Andreea, I. R. (2018). Beyond Macroeconomics of Food and Nutrition Security. *International Journal of Sustainable Economies Management, 7*(1), 13–22. doi:10.4018/IJSEM.2018010102

Anwar, J. (2017). Reproductive and Mental Health during Natural Disaster: Implications and Issues for Women in Developing Nations – A Case Example. In I. Management Association (Ed.), Gaming and Technology Addiction: Breakthroughs in Research and Practice (pp. 446-472). Hershey, PA: IGI Global. doi:10.4018/978-1-5225-0778-9.ch021

Awadh, H., Aksissou, M., Benhardouze, W., Darasi, F., & Snaiki, J. (2018). Socioeconomic Status of Artisanal Fishers in the West Part of Moroccan Mediterranean. *International Journal of Social Ecology and Sustainable Development, 9*(1), 40–52. doi:10.4018/IJSESD.2018010104

Aye, G. C., & Haruna, R. F. (2018). Effect of Climate Change on Crop Productivity and Prices in Benue State, Nigeria: Implications for Food Security. In V. Erokhin (Ed.), *Establishing Food Security and Alternatives to International Trade in Emerging Economies* (pp. 244–268). Hershey, PA: IGI Global. doi:10.4018/978-1-5225-2733-6.ch012

Azizan, S. A., & Suki, N. M. (2017). Consumers' Intentions to Purchase Organic Food Products. In T. Esakki (Ed.), *Green Marketing and Environmental Responsibility in Modern Corporations* (pp. 68–82). Hershey, PA: IGI Global. doi:10.4018/978-1-5225-2331-4.ch005

Barakabitze, A. A., Fue, K. G., Kitindi, E. J., & Sanga, C. A. (2017). Developing a Framework for Next Generation Integrated Agro Food-Advisory Systems in Developing Countries. In I. Management Association (Ed.), Agri-Food Supply Chain Management: Breakthroughs in Research and Practice (pp. 47-67). Hershey, PA: IGI Global. doi:10.4018/978-1-5225-1629-3.ch004

Beachcroft-Shaw, H., & Ellis, D. (2018). Using Successful Cases to Promote Environmental Sustainability: A Social Marketing Approach. In I. Management Association (Ed.), Sustainable Development: Concepts, Methodologies, Tools, and Applications (pp. 936-953). Hershey, PA: IGI Global. doi:10.4018/978-1-5225-3817-2.ch042

Behnassi, M., Kahime, K., Boussaa, S., Boumezzough, A., & Messouli, M. (2017). Infectious Diseases and Climate Vulnerability in Morocco: Governance and Adaptation Options. In I. Management Association (Ed.), Public Health and Welfare: Concepts, Methodologies, Tools, and Applications (pp. 91-109). Hershey, PA: IGI Global. doi:10.4018/978-1-5225-1674-3.ch005

Bekele, F., & Bekele, I. (2017). Social and Environmental Impacts on Agricultural Development. In W. Ganpat, R. Dyer, & W. Isaac (Eds.), *Agricultural Development and Food Security in Developing Nations* (pp. 21–56). Hershey, PA: IGI Global. doi:10.4018/978-1-5225-0942-4.ch002

Benaouda, A., & García-Peñalvo, F. J. (2018). Towards an Intelligent System for the Territorial Planning: Agricultural Case. In F. García-Peñalvo (Ed.), *Global Implications of Emerging Technology Trends* (pp. 158–178). Hershey, PA: IGI Global. doi:10.4018/978-1-5225-4944-4.ch010

Bhaskar, A., Rao, G. B., & Vencatesan, J. (2017). Characterization and Management Concerns of Water Resources around Pallikaranai Marsh, South Chennai. In P. Rao & Y. Patil (Eds.), *Reconsidering the Impact of Climate Change on Global Water Supply, Use, and Management* (pp. 102–121). Hershey, PA: IGI Global. doi:10.4018/978-1-5225-1046-8.ch007

Bianchini, E., & Sicurella, S. (2010). GIS: A New Tool for Criminology and Victimology's Studies. In R. Sette (Ed.), *Cases on Technologies for Teaching Criminology and Victimology: Methodologies and Practices* (pp. 87–110). Hershey, PA: IGI Global. doi:10.4018/978-1-60566-872-7.ch006

Bogataj, D., & Drobne, D. (2017). Control of Perishable Goods in Cold Logistic Chains by Bionanosensors. In S. Joo (Ed.), *Applying Nanotechnology for Environmental Sustainability* (pp. 376–402). Hershey, PA: IGI Global. doi:10.4018/978-1-5225-0585-3.ch016

Bogataj, D., & Drobne, D. (2017). Control of Perishable Goods in Cold Logistic Chains by Bionanosensors. In I. Management Association (Ed.), Materials Science and Engineering: Concepts, Methodologies, Tools, and Applications (pp. 471-497). Hershey, PA: IGI Global. doi:10.4018/978-1-5225-1798-6.ch019

Bogueva, D., & Marinova, D. (2018). What Is More Important: Perception of Masculinity or Personal Health and the Environment? In D. Bogueva, D. Marinova, & T. Raphaely (Eds.), *Handbook of Research on Social Marketing and Its Influence on Animal Origin Food Product Consumption* (pp. 148–162). Hershey, PA: IGI Global. doi:10.4018/978-1-5225-4757-0.ch010

Bouzid, M. (2017). Waterborne Diseases and Climate Change: Impact and Implications. In M. Bouzid (Ed.), *Examining the Role of Environmental Change on Emerging Infectious Diseases and Pandemics* (pp. 89–108). Hershey, PA: IGI Global. doi:10.4018/978-1-5225-0553-2.ch004

Bowles, D. C. (2017). Climate Change-Associated Conflict and Infectious Disease. In M. Bouzid (Ed.), *Examining the Role of Environmental Change on Emerging Infectious Diseases and Pandemics* (pp. 68–88). Hershey, PA: IGI Global. doi:10.4018/978-1-5225-0553-2.ch003

Buck, J. J., & Lowry, R. K. (2017). Oceanographic Data Management: Quills and Free Text to the Digital Age and "Big Data". In P. Diviacco, A. Leadbetter, & H. Glaves (Eds.), *Oceanographic and Marine Cross-Domain Data Management for Sustainable Development* (pp. 1–22). Hershey, PA: IGI Global. doi:10.4018/978-1-5225-0700-0.ch001

Buse, C. G. (2017). Are Climate Change Adaptation Policies a Game Changer?: A Case Study of Perspectives from Public Health Officials in Ontario, Canada. In M. Bouzid (Ed.), *Examining the Role of Environmental Change on Emerging Infectious Diseases and Pandemics* (pp. 230–257). Hershey, PA: IGI Global. doi:10.4018/978-1-5225-0553-2.ch010

Calderon, F. A., Giolo, E. G., Frau, C. D., Rengel, M. G., Rodriguez, H., Tornello, M., ... Gallucci, R. (2018). Seismic Microzonation and Site Effects Detection Through Microtremors Measures: A Review. In N. Ceryan (Ed.), *Handbook of Research on Trends and Digital Advances in Engineering Geology* (pp. 326–349). Hershey, PA: IGI Global. doi:10.4018/978-1-5225-2709-1.ch009

Carfì, D., Donato, A., & Panuccio, D. (2018). A Game Theory Coopetitive Perspective for Sustainability of Global Feeding: Agreements Among Vegan and Non-Vegan Food Firms. In I. Management Association (Ed.), Game Theory: Breakthroughs in Research and Practice (pp. 71-104). Hershey, PA: IGI Global. doi:10.4018/978-1-5225-2594-3.ch004

Carr, D. A., & Carr, T. R. (2007). Geographic Information System Applications in the Public Sector. In G. Garson (Ed.), *Modern Public Information Technology Systems: Issues and Challenges* (pp. 293–311). Hershey, PA: IGI Global. doi:10.4018/978-1-59904-051-6.ch015

Castagnolo, V. (2018). Analyzing, Classifying, Safeguarding: Drawing for the Borgo Murattiano Neighbourhood of Bari. In G. Carlone, N. Martinelli, & F. Rotondo (Eds.), *Designing Grid Cities for Optimized Urban Development and Planning* (pp. 93–108). Hershey, PA: IGI Global. doi:10.4018/978-1-5225-3613-0.ch006

Chekima, B. (2018). The Dilemma of Purchase Intention: A Conceptual Framework for Understanding Actual Consumption of Organic Food. *International Journal of Sustainable Economies Management*, *7*(2), 1–13. doi:10.4018/IJSEM.2018040101

Chen, Y. (2017). Sustainable Supply Chains and International Soft Landings: A Case of Wetland Entrepreneurship. In B. Christiansen & F. Kasarcı (Eds.), *Corporate Espionage, Geopolitics, and Diplomacy Issues in International Business* (pp. 232–247). Hershey, PA: IGI Global. doi:10.4018/978-1-5225-1031-4.ch013

Çıtak, L., Akel, V., & Ersoy, E. (2018). Investors' Reactions to the Announcement of New Constituents of BIST Sustainability Index: An Analysis by Event Study and Mean-Median Tests. In M. Risso & S. Testarmata (Eds.), *Value Sharing for Sustainable and Inclusive Development* (pp. 270–289). Hershey, PA: IGI Global. doi:10.4018/978-1-5225-3147-0.ch012

Crossland, M. D. (2009). Geographic Information Systems as Decision Tools. In M. Khosrow-Pour, D.B.A. (Ed.), Encyclopedia of Information Science and Technology, Second Edition (pp. 1630-1633). Hershey, PA: IGI Global. doi:10.4018/978-1-60566-026-4.ch257

Crossland, M. D., Herschel, R. T., Perkins, W. C., & Scudder, J. N. (2002). Geographic Information Systems: How Cognitive Style Impacts Decision-Making Effectiveness. In M. Mahmood (Ed.), *Advanced Topics in End User Computing* (Vol. 1, pp. 285–301). Hershey, PA: IGI Global. doi:10.4018/978-1-930708-42-6.ch017

D'Aleo, V., D'Aleo, F., & Bonanno, R. (2018). New Food Industries Toward a New Level of Sustainable Supply: Success Stories, Business Models, and Strategies. In V. Erokhin (Ed.), *Establishing Food Security and Alternatives to International Trade in Emerging Economies* (pp. 74–97). Hershey, PA: IGI Global. doi:10.4018/978-1-5225-2733-6.ch004

Dagevos, H., & Reinders, M. J. (2018). Flexitarianism and Social Marketing: Reflections on Eating Meat in Moderation. In D. Bogueva, D. Marinova, & T. Raphaely (Eds.), *Handbook of Research on Social Marketing and Its Influence on Animal Origin Food Product Consumption* (pp. 105–120). Hershey, PA: IGI Global. doi:10.4018/978-1-5225-4757-0.ch007

Deenapanray, P. N., & Ramma, I. (2017). Adaptations to Climate Change and Climate Variability in the Agriculture Sector in Mauritius: Lessons from a Technical Needs Assessment. In I. Management Association (Ed.), Natural Resources Management: Concepts, Methodologies, Tools, and Applications (pp. 655-680). Hershey, PA: IGI Global. doi:10.4018/978-1-5225-0803-8.ch030

Deenapanray, P. N., & Ramma, I. (2017). Adaptations to Climate Change and Climate Variability in the Agriculture Sector in Mauritius: Lessons from a Technical Needs Assessment. In I. Management Association (Ed.), Natural Resources Management: Concepts, Methodologies, Tools, and Applications (pp. 655-680). Hershey, PA: IGI Global. doi:10.4018/978-1-5225-0803-8.ch030

Deshpande, S., Basu, S. K., Li, X., & Chen, X. (2017). Smart, Innovative and Intelligent Technologies Used in Drug Designing. In I. Management Association (Ed.), Pharmaceutical Sciences: Breakthroughs in Research and Practice (pp. 1175-1191). Hershey, PA: IGI Global. doi:10.4018/978-1-5225-1762-7.ch045

Dlamini, P. N. (2017). Use of Information and Communication Technologies Tools to Capture, Store, and Disseminate Indigenous Knowledge: A Literature Review. In P. Ngulube (Ed.), *Handbook of Research on Theoretical Perspectives on Indigenous Knowledge Systems in Developing Countries* (pp. 225–247). Hershey, PA: IGI Global. doi:10.4018/978-1-5225-0833-5.ch010

Dolejsova, M., & Kera, D. (2017). The Fermentation GutHub Project and the Internet of Microbes. In S. Konomi & G. Roussos (Eds.), *Enriching Urban Spaces with Ambient Computing, the Internet of Things, and Smart City Design* (pp. 25–46). Hershey, PA: IGI Global. doi:10.4018/978-1-5225-0827-4.ch002

Dolunay, O. (2018). A Paradigm Shift: Empowering Farmers to Eliminate the Waste in the Form of Fresh Water and Energy Through the Implementation of 4R+T. In I. Management Association (Ed.), Sustainable Development: Concepts, Methodologies, Tools, and Applications (pp. 882-892). Hershey, PA: IGI Global. doi:10.4018/978-1-5225-3817-2.ch039

Dube, P., Heijman, W. J., Ihle, R., & Ochieng, J. (2018). The Potential of Traditional Leafy Vegetables for Improving Food Security in Africa. In V. Erokhin (Ed.), *Establishing Food Security and Alternatives to International Trade in Emerging Economies* (pp. 220–243). Hershey, PA: IGI Global. doi:10.4018/978-1-5225-2733-6.ch011

Duruji, M. M., & Urenma, D. F. (2017). The Environmentalism and Politics of Climate Change: A Study of the Process of Global Convergence through UNFCCC Conferences. In I. Management Association (Ed.), *Natural Resources Management: Concepts, Methodologies, Tools, and Applications* (pp. 77-108). Hershey, PA: IGI Global. 10.4018/978-1-5225-0803-8.ch004

Dutta, U. (2017). Agro-Geoinformatics, Potato Cultivation, and Climate Change. In S. Londhe (Ed.), *Sustainable Potato Production and the Impact of Climate Change* (pp. 247–271). Hershey, PA: IGI Global. doi:10.4018/978-1-5225-1715-3.ch012

Edirisinghe, R., Stranieri, A., & Wickramasinghe, N. (2017). A Taxonomy for mHealth. In N. Wickramasinghe (Ed.), *Handbook of Research on Healthcare Administration and Management* (pp. 596–615). Hershey, PA: IGI Global. doi:10.4018/978-1-5225-0920-2.ch036

Ekpeni, N. M., & Ayeni, A. O. (2018). Global Natural Hazard and Disaster Vulnerability Management. In A. Eneanya (Ed.), *Handbook of Research on Environmental Policies for Emergency Management and Public Safety* (pp. 83–104). Hershey, PA: IGI Global. doi:10.4018/978-1-5225-3194-4.ch005

Ene, C., Voica, M. C., & Panait, M. (2017). Green Investments and Food Security: Opportunities and Future Directions in the Context of Sustainable Development. In M. Mieila (Ed.), *Measuring Sustainable Development and Green Investments in Contemporary Economies* (pp. 163–200). Hershey, PA: IGI Global. doi:10.4018/978-1-5225-2081-8.ch007

Escamilla, I., Ruíz, M. T., Ibarra, M. M., Soto, V. L., Quintero, R., & Guzmán, G. (2018). Geocoding Tweets Based on Semantic Web and Ontologies. In M. Lytras, N. Aljohani, E. Damiani, & K. Chui (Eds.), *Innovations, Developments, and Applications of Semantic Web and Information Systems* (pp. 372–392). Hershey, PA: IGI Global. doi:10.4018/978-1-5225-5042-6.ch014

Escribano, A. J. (2018). Marketing Strategies for Trendy Animal Products: Sustainability as a Core. In F. Quoquab, R. Thurasamy, & J. Mohammad (Eds.), *Driving Green Consumerism Through Strategic Sustainability Marketing* (pp. 169–203). Hershey, PA: IGI Global. doi:10.4018/978-1-5225-2912-5.ch010

Eudoxie, G., & Roopnarine, R. (2017). Climate Change Adaptation and Disaster Risk Management in the Caribbean. In W. Ganpat & W. Isaac (Eds.), *Environmental Sustainability and Climate Change Adaptation Strategies* (pp. 97–125). Hershey, PA: IGI Global. doi:10.4018/978-1-5225-1607-1.ch004

Farmer, L. S. (2017). Data Analytics for Strategic Management: Getting the Right Data. In V. Wang (Ed.), *Encyclopedia of Strategic Leadership and Management* (pp. 810–822). Hershey, PA: IGI Global. doi:10.4018/978-1-5225-1049-9.ch056

Fattal, L. R. (2017). Catastrophe: An Uncanny Catalyst for Creativity. In R. Shin (Ed.), *Convergence of Contemporary Art, Visual Culture, and Global Civic Engagement* (pp. 244–262). Hershey, PA: IGI Global. doi:10.4018/978-1-5225-1665-1.ch014

Forti, I. (2017). A Cross Reading of Landscape through Digital Landscape Models: The Case of Southern Garda. In A. Ippolito (Ed.), *Handbook of Research on Emerging Technologies for Architectural and Archaeological Heritage* (pp. 532–561). Hershey, PA: IGI Global. doi:10.4018/978-1-5225-0675-1.ch018

Gharbi, A., De Runz, C., & Akdag, H. (2017). Urban Development Modelling: A Survey. In S. Faiz & K. Mahmoudi (Eds.), *Handbook of Research on Geographic Information Systems Applications and Advancements* (pp. 96–124). Hershey, PA: IGI Global. doi:10.4018/978-1-5225-0937-0.ch004

Ghosh, I., & Ghoshal, I. (2018). Implications of Trade Liberalization for Food Security Under the ASEAN-India Strategic Partnership: A Gravity Model Approach. In V. Erokhin (Ed.), *Establishing Food Security and Alternatives to International Trade in Emerging Economies* (pp. 98–118). Hershey, PA: IGI Global. doi:10.4018/978-1-5225-2733-6.ch005

Glaves, H. M. (2017). Developing a Common Global Framework for Marine Data Management. In P. Diviacco, A. Leadbetter, & H. Glaves (Eds.), *Oceanographic and Marine Cross-Domain Data Management for Sustainable Development* (pp. 47–68). Hershey, PA: IGI Global. doi:10.4018/978-1-5225-0700-0.ch003

Godulla, A., & Wolf, C. (2018). Future of Food: Transmedia Strategies of National Geographic. In R. Gambarato & G. Alzamora (Eds.), *Exploring Transmedia Journalism in the Digital Age* (pp. 162–182). Hershey, PA: IGI Global. doi:10.4018/978-1-5225-3781-6.ch010

Gomes, P. P. (2018). Food and Environment: A Review on the Sustainability of Six Different Dietary Patterns. In A. Obayelu (Ed.), *Food Systems Sustainability and Environmental Policies in Modern Economies* (pp. 15–31). Hershey, PA: IGI Global. doi:10.4018/978-1-5225-3631-4.ch002

Gonzalez-Feliu, J. (2018). Sustainability Evaluation of Green Urban Logistics Systems: Literature Overview and Proposed Framework. In A. Paul, D. Bhattacharyya, & S. Anand (Eds.), *Green Initiatives for Business Sustainability and Value Creation* (pp. 103–134). Hershey, PA: IGI Global. doi:10.4018/978-1-5225-2662-9.ch005

Goodland, R. (2017). A Fresh Look at Livestock Greenhouse Gas Emissions and Mitigation. In I. Management Association (Ed.), Natural Resources Management: Concepts, Methodologies, Tools, and Applications (pp. 124-139). Hershey, PA: IGI Global. doi:10.4018/978-1-5225-0803-8.ch006

Goundar, S., & Appana, S. (2018). Mainstreaming Development Policies for Climate Change in Fiji: A Policy Gap Analysis and the Role of ICTs. In I. Management Association (Ed.), Sustainable Development: Concepts, Methodologies, Tools, and Applications (pp. 402-432). Hershey, PA: IGI Global. doi:10.4018/978-1-5225-3817-2.ch020

Granell-Canut, C., & Aguilar-Moreno, E. (2018). Geospatial Influence in Science Mapping. In M. Khosrow-Pour, D.B.A. (Ed.), Encyclopedia of Information Science and Technology, Fourth Edition (pp. 3473-3483). Hershey, PA: IGI Global. doi:10.4018/978-1-5225-2255-3.ch302

Grigelis, A., Blažauskas, N., Gelumbauskaitė, L. Ž., Gulbinskas, S., Suzdalev, S., & Ferrarin, C. (2017). Marine Environment Data Management Related to the Human Activity in the South-Eastern Baltic Sea (The Lithuanian Sector). In P. Diviacco, A. Leadbetter, & H. Glaves (Eds.), *Oceanographic and Marine Cross-Domain Data Management for Sustainable Development* (pp. 282–302). Hershey, PA: IGI Global. doi:10.4018/978-1-5225-0700-0.ch012

Guma, I. P., Rwashana, A. S., & Oyo, B. (2018). Food Security Policy Analysis Using System Dynamics: The Case of Uganda. *International Journal of Information Technologies and Systems Approach*, *11*(1), 72–90. doi:10.4018/IJITSA.2018010104

Guma, I. P., Rwashana, A. S., & Oyo, B. (2018). Food Security Indicators for Subsistence Farmers Sustainability: A System Dynamics Approach. *International Journal of System Dynamics Applications*, *7*(1), 45–64. doi:10.4018/IJSDA.2018010103

Gupta, P., & Goyal, S. (2017). Wildlife Habitat Evaluation. In A. Santra & S. Mitra (Eds.), *Remote Sensing Techniques and GIS Applications in Earth and Environmental Studies* (pp. 258–264). Hershey, PA: IGI Global. doi:10.4018/978-1-5225-1814-3.ch013

Hanson, T., & Hildebrand, E. (2018). GPS Travel Diaries in Rural Transportation Research: A Focus on Older Drivers. In I. Management Association (Ed.), Intelligent Transportation and Planning: Breakthroughs in Research and Practice (pp. 609-625). Hershey, PA: IGI Global. doi:10.4018/978-1-5225-5210-9.ch027

Hartman, M. B. (2017). Research-Based Climate Change Public Education Programs. In I. Management Association (Ed.), Natural Resources Management: Concepts, Methodologies, Tools, and Applications (pp. 992-1003). Hershey, PA: IGI Global. doi:10.4018/978-1-5225-0803-8.ch046

Hashim, N. (2017). Zanzibari Seaweed: Global Climate Change and the Promise of Adaptation. In I. Management Association (Ed.), Natural Resources Management: Concepts, Methodologies, Tools, and Applications (pp. 365-391). Hershey, PA: IGI Global. doi:10.4018/978-1-5225-0803-8.ch019

Herrera, J. E., Argüello, L. V., Gonzalez-Feliu, J., & Jaimes, W. A. (2017). Decision Support System Design Requirements, Information Management, and Urban Logistics Efficiency: Case Study of Bogotá, Colombia. In G. Jamil, A. Soares, & C. Pessoa (Eds.), *Handbook of Research on Information Management for Effective Logistics and Supply Chains* (pp. 223–238). Hershey, PA: IGI Global. doi:10.4018/978-1-5225-0973-8.ch012

Huizinga, T., Ayanso, A., Smoor, M., & Wronski, T. (2017). Exploring Insurance and Natural Disaster Tweets Using Text Analytics. *International Journal of Business Analytics*, *4*(1), 1–17. doi:10.4018/IJBAN.2017010101

Hung, K., Kalantari, M., & Rajabifard, A. (2017). An Integrated Method for Assessing the Text Content Quality of Volunteered Geographic Information in Disaster Management. *International Journal of Information Systems for Crisis Response and Management*, *9*(2), 1–17. doi:10.4018/IJISCRAM.2017040101

Husnain, A., & Avdic, A. (2018). Identifying the Contemporary Status of E-Service Sustainability Research. In I. Management Association (Ed.), Sustainable Development: Concepts, Methodologies, Tools, and Applications (pp. 467-485). Hershey, PA: IGI Global. doi:10.4018/978-1-5225-3817-2.ch022

Iarossi, M. P., & Ferro, L. (2017). "The Past is Never Dead. It's Not Even Past": Virtual Archaeological Promenade. In A. Ippolito & M. Cigola (Eds.), *Handbook of Research on Emerging Technologies for Digital Preservation and Information Modeling* (pp. 228–255). Hershey, PA: IGI Global. doi:10.4018/978-1-5225-0680-5.ch010

Ignjatijević, S., & Cvijanović, D. (2018). Analysis of Serbian Honey Production and Exports. In *Exploring the Global Competitiveness of Agri-Food Sectors and Serbia's Dominant Presence: Emerging Research and Opportunities* (pp. 109–139). Hershey, PA: IGI Global. doi:10.4018/978-1-5225-2762-6.ch005

Jana, S. K., & Karmakar, A. K. (2017). Globalization, Governance, and Food Security: The Case of BRICS. In I. Management Association (Ed.), Natural Resources Management: Concepts, Methodologies, Tools, and Applications (pp. 692-712). Hershey, PA: IGI Global. doi:10.4018/978-1-5225-0803-8.ch032

Jana, S. K., & Karmakar, A. K. (2017). Food Security in Asia: Is There Convergence? In I. Management Association (Ed.), Natural Resources Management: Concepts, Methodologies, Tools, and Applications (pp. 109-123). Hershey, PA: IGI Global. doi:10.4018/978-1-5225-0803-8.ch005

John, J., & Kumar, S. (2018). A Locational Decision Making Framework for Shipbreaking Under Multiple Criteria. In I. Management Association (Ed.), Operations and Service Management: Concepts, Methodologies, Tools, and Applications (pp. 504-527). Hershey, PA: IGI Global. doi:10.4018/978-1-5225-3909-4.ch024

John, J., & Srivastava, R. K. (2018). Decision Insights for Shipbreaking using Environmental Impact Assessment: Review and Perspectives. *International Journal of Strategic Decision Sciences*, *9*(1), 45–62. doi:10.4018/IJSDS.2018010104

Joshi, Y., & Rahman, Z. (2018). Determinants of Sustainable Consumption Behaviour: Review and Conceptual Framework. In A. Paul, D. Bhattacharyya, & S. Anand (Eds.), *Green Initiatives for Business Sustainability and Value Creation* (pp. 239–262). Hershey, PA: IGI Global. doi:10.4018/978-1-5225-2662-9.ch011

Juma, D. W., Reuben, M., Wang, H., & Li, F. (2018). Adaptive Coevolution: Realigning the Water Governance Regime to the Changing Climate. In I. Management Association (Ed.), Hydrology and Water Resource Management: Breakthroughs in Research and Practice (pp. 346-357). Hershey, PA: IGI Global. doi:10.4018/978-1-5225-3427-3.ch014

K., S., & Tripathy, B. K. (2018). Neighborhood Rough-Sets-Based Spatial Data Analytics. In M. Khosrow-Pour, D.B.A. (Ed.), *Encyclopedia of Information Science and Technology, Fourth Edition* (pp. 1835-1844). Hershey, PA: IGI Global. doi:10.4018/978-1-5225-2255-3.ch160

Kabir, F. (2018). Towards a More Gender-Inclusive Climate Change Policy. In N. Mahtab, T. Haque, I. Khan, M. Islam, & I. Wahid (Eds.), *Handbook of Research on Women's Issues and Rights in the Developing World* (pp. 354–369). Hershey, PA: IGI Global. doi:10.4018/978-1-5225-3018-3.ch022

Kabir, F. (2018). Towards a More Gender-Inclusive Climate Change Policy. In I. Management Association (Ed.), Climate Change and Environmental Concerns: Breakthroughs in Research and Practice (pp. 525-540). Hershey, PA: IGI Global. doi:10.4018/978-1-5225-5487-5.ch027

Kanyamuka, J. S., Jumbe, C. B., & Ricker-Gilbert, J. (2018). Making Agricultural Input Subsidies More Effective and Profitable in Africa: The Role of Complementary Interventions. In A. Obayelu (Ed.), *Food Systems Sustainability and Environmental Policies in Modern Economies* (pp. 172–187). Hershey, PA: IGI Global. doi:10.4018/978-1-5225-3631-4.ch008

Karimi, H., & Gholamrezafahimi, F. (2017). Study of Integrated Coastal Zone Management and Its Environmental Effects: A Case of Iran. In R. Singh, A. Singh, & V. Srivastava (Eds.), *Environmental Issues Surrounding Human Overpopulation* (pp. 64–88). Hershey, PA: IGI Global. doi:10.4018/978-1-5225-1683-5.ch004

Kaya, I. R., Hutabarat, J., & Bambang, A. N. (2018). "Sasi": A New Path to Sustain Seaweed Farming From Up-Stream to Down-Stream in Kotania Bay, Molucass. *International Journal of Social Ecology and Sustainable Development, 9*(2), 28–36. doi:10.4018/IJSESD.2018040103

Khader, V. (2018). Technologies for Food, Health, Livelihood, and Nutrition Security. In I. Management Association (Ed.), Food Science and Nutrition: Breakthroughs in Research and Practice (pp. 94-112). Hershey, PA: IGI Global. doi:10.4018/978-1-5225-5207-9.ch005

Kocadağlı, A. Y. (2017). The Temporal and Spatial Development of Organic Agriculture in Turkey. In W. Ganpat, R. Dyer, & W. Isaac (Eds.), *Agricultural Development and Food Security in Developing Nations* (pp. 130–156). Hershey, PA: IGI Global. doi:10.4018/978-1-5225-0942-4.ch006

Koundouri, P., Giannouli, A., & Souliotis, I. (2017). An Integrated Approach for Sustainable Environmental and Socio-Economic Development Using Offshore Infrastructure. In I. Management Association (Ed.), Renewable and Alternative Energy: Concepts, Methodologies, Tools, and Applications (pp. 1581-1601). Hershey, PA: IGI Global. doi:10.4018/978-1-5225-1671-2.ch056

Kumar, A., & Dash, M. K. (2017). Sustainability and Future Generation Infrastructure on Digital Platform: A Study of Generation Y. In N. Ray (Ed.), *Business Infrastructure for Sustainability in Developing Economies* (pp. 124–142). Hershey, PA: IGI Global. doi:10.4018/978-1-5225-2041-2.ch007

Kumar, A., Mukherjee, A. B., & Krishna, A. P. (2017). Application of Conventional Data Mining Techniques and Web Mining to Aid Disaster Management. In A. Kumar (Ed.), *Web Usage Mining Techniques and Applications Across Industries* (pp. 138–167). Hershey, PA: IGI Global. doi:10.4018/978-1-5225-0613-3.ch006

Kumar, C. P. (2017). Impact of Climate Change on Groundwater Resources. In I. Management Association (Ed.), Natural Resources Management: Concepts, Methodologies, Tools, and Applications (pp. 1094-1120). Hershey, PA: IGI Global. doi:10.4018/978-1-5225-0803-8.ch052

Kumari, S., & Patil, Y. (2017). Achieving Climate Smart Agriculture with a Sustainable Use of Water: A Conceptual Framework for Sustaining the Use of Water for Agriculture in the Era of Climate Change. In P. Rao & Y. Patil (Eds.), *Reconsidering the Impact of Climate Change on Global Water Supply, Use, and Management* (pp. 122–143). Hershey, PA: IGI Global. doi:10.4018/978-1-5225-1046-8.ch008

Kumari, S., & Patil, Y. (2018). Achieving Climate Smart Agriculture With a Sustainable Use of Water: A Conceptual Framework for Sustaining the Use of Water for Agriculture in the Era of Climate Change. In I. Management Association (Ed.), Climate Change and Environmental Concerns: Breakthroughs in Research and Practice (pp. 111-133). Hershey, PA: IGI Global. doi:10.4018/978-1-5225-5487-5.ch006

Kursah, M. B. (2017). Least-Cost Pipeline using Geographic Information System: The Limit to Technicalities. *International Journal of Applied Geospatial Research, 8*(3), 1–15. doi:10.4018/ijagr.2017070101

Lahiri, S., Ghosh, D., & Bhakta, J. N. (2017). Role of Microbes in Eco-Remediation of Perturbed Aquatic Ecosystem. In J. Bhakta (Ed.), *Handbook of Research on Inventive Bioremediation Techniques* (pp. 70–107). Hershey, PA: IGI Global. doi:10.4018/978-1-5225-2325-3.ch004

Lallo, C. H., Smalling, S., Facey, A., & Hughes, M. (2018). The Impact of Climate Change on Small Ruminant Performance in Caribbean Communities. In I. Management Association (Ed.), Climate Change and Environmental Concerns: Breakthroughs in Research and Practice (pp. 193-218). Hershey, PA: IGI Global. doi:10.4018/978-1-5225-5487-5.ch010

Laurini, R. (2017). Nature of Geographic Knowledge Bases. In S. Faiz & K. Mahmoudi (Eds.), *Handbook of Research on Geographic Information Systems Applications and Advancements* (pp. 29–60). Hershey, PA: IGI Global. doi:10.4018/978-1-5225-0937-0.ch002

Lawrence, J., Simpson, L., & Piggott, A. (2017). Protected Agriculture: A Climate Change Adaptation for Food and Nutrition Security. In I. Management Association (Ed.), Natural Resources Management: Concepts, Methodologies, Tools, and Applications (pp. 140-158). Hershey, PA: IGI Global. doi:10.4018/978-1-5225-0803-8.ch007

Leadbetter, A., Cheatham, M., Shepherd, A., & Thomas, R. (2017). Linked Ocean Data 2.0. In P. Diviacco, A. Leadbetter, & H. Glaves (Eds.), *Oceanographic and Marine Cross-Domain Data Management for Sustainable Development* (pp. 69–99). Hershey, PA: IGI Global. doi:10.4018/978-1-5225-0700-0.ch004

Leipnik, M. R., & Mehta, S. S. (2005). Geographic Information Systems (GIS) in E-Marketing. In I. Clarke III & T. Flaherty (Eds.), *Advances in Electronic Marketing* (pp. 193–210). Hershey, PA: IGI Global. doi:10.4018/978-1-59140-321-0.ch011

Lucas, M. R., Rego, C., Vieira, C., & Vieira, I. (2017). Proximity and Cooperation for Innovative Regional Development: The Case of the Science and Technology Park of Alentejo. In L. Carvalho (Ed.), *Handbook of Research on Entrepreneurial Development and Innovation Within Smart Cities* (pp. 199–228). Hershey, PA: IGI Global. doi:10.4018/978-1-5225-1978-2.ch010

Ma, X., Beaulieu, S. E., Fu, L., Fox, P., Di Stefano, M., & West, P. (2017). Documenting Provenance for Reproducible Marine Ecosystem Assessment in Open Science. In P. Diviacco, A. Leadbetter, & H. Glaves (Eds.), *Oceanographic and Marine Cross-Domain Data Management for Sustainable Development* (pp. 100–126). Hershey, PA: IGI Global. doi:10.4018/978-1-5225-0700-0.ch005

Mabe, L. K., & Oladele, O. I. (2017). Application of Information Communication Technologies for Agricultural Development through Extension Services: A Review. In T. Tossy (Ed.), *Information Technology Integration for Socio-Economic Development* (pp. 52–101). Hershey, PA: IGI Global. doi:10.4018/978-1-5225-0539-6.ch003

Malomo, B. I. (2018). A Review of Psychological Resilience as a Response to Natural Hazards in Nigeria. In A. Eneanya (Ed.), *Handbook of Research on Environmental Policies for Emergency Management and Public Safety* (pp. 147–165). Hershey, PA: IGI Global. doi:10.4018/978-1-5225-3194-4.ch008

Manchiraju, S. (2018). Predicting Behavioral Intentions Toward Sustainable Fashion Consumption: A Comparison of Attitude-Behavior and Value-Behavior Consistency Models. In I. Management Association (Ed.), Fashion and Textiles: Breakthroughs in Research and Practice (pp. 1-21). Hershey, PA: IGI Global. doi:10.4018/978-1-5225-3432-7.ch001

Manzella, G. M., Bartolini, R., Bustaffa, F., D'Angelo, P., De Mattei, M., Frontini, F., ... Spada, A. (2017). Semantic Search Engine for Data Management and Sustainable Development: Marine Planning Service Platform. In P. Diviacco, A. Leadbetter, & H. Glaves (Eds.), *Oceanographic and Marine Cross-Domain Data Management for Sustainable Development* (pp. 127–154). Hershey, PA: IGI Global. doi:10.4018/978-1-5225-0700-0.ch006

Mbonigaba, J. (2018). Comparing the Effects of Unsustainable Production and Consumption of Food on Health and Policy Across Developed and Less Developed Countries. In A. Obayelu (Ed.), *Food Systems Sustainability and Environmental Policies in Modern Economies* (pp. 124–158). Hershey, PA: IGI Global. doi:10.4018/978-1-5225-3631-4.ch006

McKeown, A. E. (2017). Nurses, Healthcare, and Environmental Pollution and Solutions: Breaking the Cycle of Harm. In I. Management Association (Ed.), Natural Resources Management: Concepts, Methodologies, Tools, and Applications (pp. 392-415). Hershey, PA: IGI Global. doi:10.4018/978-1-5225-0803-8.ch020

Mennecke, B. E., & West, L. A. Jr. (2001). Geographic Information Systems in Developing Countries: Issues in Data Collection, Implementation and Management. *Journal of Global Information Management, 9*(4), 44–54. doi:10.4018/jgim.2001100103

Mili, B., Barua, A., & Katyaini, S. (2017). Climate Change and Adaptation through the Lens of Capability Approach: A Case Study from Darjeeling, Eastern Himalaya. In I. Management Association (Ed.), Natural Resources Management: Concepts, Methodologies, Tools, and Applications (pp. 1351-1365). Hershey, PA: IGI Global. doi:10.4018/978-1-5225-0803-8.ch064

Mir, S. A., Shah, M. A., Mir, M. M., & Iqbal, U. (2017). New Horizons of Nanotechnology in Agriculture and Food Processing Industry. In B. Nayak, A. Nanda, & M. Bhat (Eds.), *Integrating Biologically-Inspired Nanotechnology into Medical Practice* (pp. 230–258). Hershey, PA: IGI Global. doi:10.4018/978-1-5225-0610-2.ch009

Mujere, N., & Moyce, W. (2017). Climate Change Impacts on Surface Water Quality. In W. Ganpat & W. Isaac (Eds.), *Environmental Sustainability and Climate Change Adaptation Strategies* (pp. 322–340). Hershey, PA: IGI Global. doi:10.4018/978-1-5225-1607-1.ch012

Mukherjee, A. B., Krishna, A. P., & Patel, N. (2018). Geospatial Technology for Urban Sciences. In *Geospatial Technologies in Urban System Development: Emerging Research and Opportunities* (pp. 99–120). Hershey, PA: IGI Global. doi:10.4018/978-1-5225-3683-3.ch005

Nagarajan, S. K., & Sangaiah, A. K. (2017). Vegetation Index: Ideas, Methods, Influences, and Trends. In N. Kumar, A. Sangaiah, M. Arun, & S. Anand (Eds.), *Advanced Image Processing Techniques and Applications* (pp. 347–386). Hershey, PA: IGI Global. doi:10.4018/978-1-5225-2053-5.ch016

Naraine, L., & Meehan, K. (2017). Strengthening Food Security with Sustainable Practices by Smallholder Farmers in Lesser Developed Economies. In W. Ganpat, R. Dyer, & W. Isaac (Eds.), *Agricultural Development and Food Security in Developing Nations* (pp. 57–81). Hershey, PA: IGI Global. doi:10.4018/978-1-5225-0942-4.ch003

Naraine, L., & Meehan, K. (2017). Strengthening Food Security with Sustainable Practices by Smallholder Farmers in Lesser Developed Economies. In W. Ganpat, R. Dyer, & W. Isaac (Eds.), *Agricultural Development and Food Security in Developing Nations* (pp. 57–81). Hershey, PA: IGI Global. doi:10.4018/978-1-5225-0942-4.ch003

Naraine, L., & Meehan, K. (2017). Strengthening Food Security with Sustainable Practices by Smallholder Farmers in Lesser Developed Economies. In W. Ganpat, R. Dyer, & W. Isaac (Eds.), *Agricultural Development and Food Security in Developing Nations* (pp. 57–81). Hershey, PA: IGI Global. doi:10.4018/978-1-5225-0942-4.ch003

Neufeld, D. J., & Griffith, S. (2006). Isobord's Geographic Information System (GIS) Solution. In M. Khosrow-Pour, D.B.A. (Ed.), Cases on Database Technologies and Applications (pp. 64-80). Hershey, PA: IGI Global. doi:10.4018/978-1-59904-399-9.ch004

Nikolaou, K., Tsakiridou, E., Anastasiadis, F., & Mattas, K. (2017). Exploring Alternative Distribution Channels of Agricultural Products. *International Journal of Food and Beverage Manufacturing and Business Models*, *2*(2), 36–66. doi:10.4018/IJFBMBM.2017070103

Nishat, K. J., & Rahman, M. S. (2018). Disaster, Vulnerability, and Violence Against Women: Global Findings and a Research Agenda for Bangladesh. In N. Mahtab, T. Haque, I. Khan, M. Islam, & I. Wahid (Eds.), *Handbook of Research on Women's Issues and Rights in the Developing World* (pp. 235–250). Hershey, PA: IGI Global. doi:10.4018/978-1-5225-3018-3.ch014

O'Hara, S., Jones, D., & Trobman, H. B. (2018). Building an Urban Food System Through UDC Food Hubs. In A. Burtin, J. Fleming, & P. Hampton-Garland (Eds.), *Changing Urban Landscapes Through Public Higher Education* (pp. 116–143). Hershey, PA: IGI Global. doi:10.4018/978-1-5225-3454-9.ch006

Obayelu, A. E. (2018). Integrating Environment, Food Systems, and Sustainability in Feeding the Growing Population in Developing Countries. In A. Obayelu (Ed.), *Food Systems Sustainability and Environmental Policies in Modern Economies* (pp. 1–14). Hershey, PA: IGI Global. doi:10.4018/978-1-5225-3631-4.ch001

Othman, R., Nath, N., & Laswad, F. (2018). Environmental Reporting and Accounting: Sustainability Hybridisation. In G. Azevedo, J. da Silva Oliveira, R. Marques, & A. Ferreira (Eds.), *Handbook of Research on Modernization and Accountability in Public Sector Management* (pp. 130–158). Hershey, PA: IGI Global. doi:10.4018/978-1-5225-3731-1.ch007

Padigala, B. S. (2018). Traditional Water Management System for Climate Change Adaptation in Mountain Ecosystems. In I. Management Association (Ed.), Climate Change and Environmental Concerns: Breakthroughs in Research and Practice (pp. 630-655). Hershey, PA: IGI Global. doi:10.4018/978-1-5225-5487-5.ch033

Panda, C. K. (2018). Mobile Phone Usage in Agricultural Extension in India: The Current and Future Perspective. In F. Mtenzi, G. Oreku, D. Lupiana, & J. Yonazi (Eds.), *Mobile Technologies and Socio-Economic Development in Emerging Nations* (pp. 1–21). Hershey, PA: IGI Global. doi:10.4018/978-1-5225-4029-8.ch001

Pandian, S. L., Yarrakula, K., & Chaudhury, P. (2018). GIS-Based Decision Support System for Village Level: A Case Study in Andhra Pradesh. In S. Chaudhuri & N. Ray (Eds.), *GIS Applications in the Tourism and Hospitality Industry* (pp. 275–295). Hershey, PA: IGI Global. doi:10.4018/978-1-5225-5088-4.ch012

Quaranta, G., & Salvia, R. (2017). Social-Based Product Innovation and Governance in The Milk Sector: The Case of Carciocacio and Innonatura. In T. Tarnanidis, M. Vlachopoulou, & J. Papathanasiou (Eds.), *Driving Agribusiness With Technology Innovations* (pp. 293–310). Hershey, PA: IGI Global. doi:10.4018/978-1-5225-2107-5.ch015

Rahman, M. K., Schmidlin, T. W., Munro-Stasiuk, M. J., & Curtis, A. (2017). Geospatial Analysis of Land Loss, Land Cover Change, and Landuse Patterns of Kutubdia Island, Bangladesh. *International Journal of Applied Geospatial Research*, *8*(2), 45–60. doi:10.4018/IJAGR.2017040104

Rajack-Talley, T. A. (2017). Agriculture, Trade Liberalization and Poverty in the ACP Countries. In W. Ganpat, R. Dyer, & W. Isaac (Eds.), *Agricultural Development and Food Security in Developing Nations* (pp. 1–20). Hershey, PA: IGI Global. doi:10.4018/978-1-5225-0942-4.ch001

Rajamanickam, S. (2018). Exploring Landscapes in Regional Convergence: Environment and Sustainable Development in South Asia. In I. Management Association (Ed.), Sustainable Development: Concepts, Methodologies, Tools, and Applications (pp. 1051-1087). Hershey, PA: IGI Global. doi:10.4018/978-1-5225-3817-2.ch047

Rizvi, S. M., & Dearden, A. (2018). KHETI: ICT Solution for Agriculture Extension and Its Replication in Open and Distance Learning. In U. Pandey & V. Indrakanti (Eds.), *Open and Distance Learning Initiatives for Sustainable Development* (pp. 163–174). Hershey, PA: IGI Global. doi:10.4018/978-1-5225-2621-6.ch008

Roşu, L., & Macarov, L. I. (2017). Management of Drought and Floods in the Dobrogea Region. In I. Management Association (Ed.), Agri-Food Supply Chain Management: Breakthroughs in Research and Practice (pp. 372-403). Hershey, PA: IGI Global. doi:10.4018/978-1-5225-1629-3.ch016

Rouzbehani, K., & Rouzbehani, S. (2018). Mapping Women's World: GIS and the Case of Breast Cancer in the US. *International Journal of Public Health Management and Ethics*, *3*(1), 14–25. doi:10.4018/IJPHME.2018010102

Roy, D. (2018). Success Factors of Adoption of Mobile Applications in Rural India: Effect of Service Characteristics on Conceptual Model. In M. Khosrow-Pour, D.B.A. (Ed.), Green Computing Strategies for Competitive Advantage and Business Sustainability (pp. 211-238). Hershey, PA: IGI Global. doi:10.4018/978-1-5225-5017-4.ch010

Sajeva, M., Lemon, M., & Sahota, P. S. (2017). Governance for Food Security: A Framework for Social Learning and Scenario Building. *International Journal of Food and Beverage Manufacturing and Business Models*, *2*(2), 67–84. doi:10.4018/IJFBMBM.2017070104

Sallam, G. A., Youssef, T., Embaby, M. E., & Shaltot, F. (2011). Using Geographic Information System to Infollow the Fertilizers Pollution Migration. In I. Management Association (Ed.), Green Technologies: Concepts, Methodologies, Tools and Applications (pp. 564-586). Hershey, PA: IGI Global. doi:10.4018/978-1-60960-472-1.ch312

Sambhanthan, A., & Potdar, V. (2017). A Study of the Parameters Impacting Sustainability in Information Technology Organizations. *International Journal of Knowledge-Based Organizations*, *7*(3), 27–39. doi:10.4018/IJKBO.2017070103

Sanga, C., Kalungwizi, V. J., & Msuya, C. P. (2017). Bridging Gender Gaps in Provision of Agricultural Extension Service Using ICT: Experiences from Sokoine University of Agriculture (SUA) Farmer Voice Radio (FVR) Project in Tanzania. In I. Management Association (Ed.), Discrimination and Diversity: Concepts, Methodologies, Tools, and Applications (pp. 682-697). Hershey, PA: IGI Global. doi:10.4018/978-1-5225-1933-1.ch031

Santra, A., & Mitra, D. (2017). Role of Remote Sensing in Potential Fishing Zone Forecast. In A. Santra & S. Mitra (Eds.), *Remote Sensing Techniques and GIS Applications in Earth and Environmental Studies* (pp. 243–257). Hershey, PA: IGI Global. doi:10.4018/978-1-5225-1814-3.ch012

Schaap, D. (2017). SeaDataNet: Towards a Pan-European Infrastructure for Marine and Ocean Data Management. In P. Diviacco, A. Leadbetter, & H. Glaves (Eds.), *Oceanographic and Marine Cross-Domain Data Management for Sustainable Development* (pp. 155–177). Hershey, PA: IGI Global. doi:10.4018/978-1-5225-0700-0.ch007

Seckin-Celik, T. (2017). Sustainability Reporting and Sustainability in the Turkish Business Context. In U. Akkucuk (Ed.), *Ethics and Sustainability in Global Supply Chain Management* (pp. 115–132). Hershey, PA: IGI Global. doi:10.4018/978-1-5225-2036-8.ch006

Segbefia, A. Y., Barnes, V. R., Akpalu, L. A., & Mensah, M. (2018). Environmental Location Assessment for Seaweed Cultivation in Ghana: A Spatial Multi-Criteria Approach. *International Journal of Applied Geospatial Research, 9*(1), 51–64. doi:10.4018/IJAGR.2018010104

Sen, Y. (2018). How to Manage Sustainability: A Framework for Corporate Sustainability Tools. In I. Management Association (Ed.), Sustainable Development: Concepts, Methodologies, Tools, and Applications (pp. 568-589). Hershey, PA: IGI Global. doi:10.4018/978-1-5225-3817-2.ch026

Shamshiry, E., Abdulai, A. M., Mokhtar, M. B., & Komoo, I. (2015). Regional Landfill Site Selection with GIS and Analytical Hierarchy Process Techniques: A Case Study of Langkawi Island, Malaysia. In P. Thomas, M. Srihari, & S. Kaur (Eds.), *Handbook of Research on Cultural and Economic Impacts of the Information Society* (pp. 248–282). Hershey, PA: IGI Global. doi:10.4018/978-1-4666-8598-7.ch011

Sharma, Y. K., Mangla, S. K., Patil, P. P., & Uniyal, S. (2018). Analyzing Sustainable Food Supply Chain Management Challenges in India. In M. Ram & J. Davim (Eds.), *Soft Computing Techniques and Applications in Mechanical Engineering* (pp. 162–180). Hershey, PA: IGI Global. doi:10.4018/978-1-5225-3035-0.ch008

Silvestrelli, P. (2018). The Impact of Events: To Which Extent Are They Sustainable for Tourist Destinations? Some Evidences From Expo Milano 2015. In M. Risso & S. Testarmata (Eds.), *Value Sharing for Sustainable and Inclusive Development* (pp. 185–204). Hershey, PA: IGI Global. doi:10.4018/978-1-5225-3147-0.ch008

Silvius, G. (2018). Sustainability Evaluation of IT/IS Projects. In I. Management Association (Ed.), Sustainable Development: Concepts, Methodologies, Tools, and Applications (pp. 26-40). Hershey, PA: IGI Global. doi:10.4018/978-1-5225-3817-2.ch002

Singh, R., Srivastava, P., Singh, P., Upadhyay, S., & Raghubanshi, A. S. (2017). Human Overpopulation and Food Security: Challenges for the Agriculture Sustainability. In R. Singh, A. Singh, & V. Srivastava (Eds.), *Environmental Issues Surrounding Human Overpopulation* (pp. 12–39). Hershey, PA: IGI Global. doi:10.4018/978-1-5225-1683-5.ch002

Singh, R., Srivastava, P., Singh, P., Upadhyay, S., & Raghubanshi, A. S. (2017). Human Overpopulation and Food Security: Challenges for the Agriculture Sustainability. In R. Singh, A. Singh, & V. Srivastava (Eds.), *Environmental Issues Surrounding Human Overpopulation* (pp. 12–39). Hershey, PA: IGI Global. doi:10.4018/978-1-5225-1683-5.ch002

Srivastava, N. (2017). Climate Change Mitigation: Collective Efforts and Responsibly. In I. Management Association (Ed.), Natural Resources Management: Concepts, Methodologies, Tools, and Applications (pp. 64-76). Hershey, PA: IGI Global. doi:10.4018/978-1-5225-0803-8.ch003

Stanganelli, M., & Gerundo, C. (2017). Understanding the Role of Urban Morphology and Green Areas Configuration During Heat Waves. *International Journal of Agricultural and Environmental Information Systems*, *8*(2), 50–64. doi:10.4018/IJAEIS.2017040104

Stewart, M. K., Hagood, D., & Ching, C. C. (2017). Virtual Games and Real-World Communities: Environments that Constrain and Enable Physical Activity in Games for Health. *International Journal of Game-Based Learning*, *7*(1), 1–19. doi:10.4018/IJGBL.2017010101

Stone, R. J. (2017). Modelling the Frequency of Tropical Cyclones in the Lower Caribbean Region. In W. Ganpat & W. Isaac (Eds.), *Environmental Sustainability and Climate Change Adaptation Strategies* (pp. 341–349). Hershey, PA: IGI Global. doi:10.4018/978-1-5225-1607-1.ch013

Syed, A., & Jabeen, U. A. (2018). Climate Change Impact on Agriculture and Food Security. In A. Eneanya (Ed.), *Handbook of Research on Environmental Policies for Emergency Management and Public Safety* (pp. 223–237). Hershey, PA: IGI Global. doi:10.4018/978-1-5225-3194-4.ch012

Tam, G. C. (2017). The Global View of Sustainability. In *Managerial Strategies and Green Solutions for Project Sustainability* (pp. 1–24). Hershey, PA: IGI Global. doi:10.4018/978-1-5225-2371-0.ch001

Tam, G. C. (2017). Understanding Project Sustainability. In *Managerial Strategies and Green Solutions for Project Sustainability* (pp. 110–139). Hershey, PA: IGI Global. doi:10.4018/978-1-5225-2371-0.ch005

Tam, G. C. (2017). Perspectives on Sustainability. In *Managerial Strategies and Green Solutions for Project Sustainability* (pp. 53–76). Hershey, PA: IGI Global. doi:10.4018/978-1-5225-2371-0.ch003

Tang, M., & Karunanithi, A. T. (2018). Visual Logic Maps (vLms). In *Advanced Concept Maps in STEM Education: Emerging Research and Opportunities* (pp. 108–149). Hershey, PA: IGI Global. doi:10.4018/978-1-5225-2184-6.ch005

Taşçıoğlu, M., & Yener, D. (2018). The Value and Scope of GIS in Marketing and Tourism Management. In S. Chaudhuri & N. Ray (Eds.), *GIS Applications in the Tourism and Hospitality Industry* (pp. 189–211). Hershey, PA: IGI Global. doi:10.4018/978-1-5225-5088-4.ch009

Tianming, G. (2018). Food Security and Rural Development on Emerging Markets of Northeast Asia: Cases of Chinese North and Russian Far East. In V. Erokhin (Ed.), *Establishing Food Security and Alternatives to International Trade in Emerging Economies* (pp. 155–176). Hershey, PA: IGI Global. doi:10.4018/978-1-5225-2733-6.ch008

Tiftikçigil, B. Y., Yaşgül, Y. S., & Güriş, B. (2017). Sustainability of Foreign Trade Deficit in Energy: The Case of Turkey. In N. Ray (Ed.), *Business Infrastructure for Sustainability in Developing Economies* (pp. 94–109). Hershey, PA: IGI Global. doi:10.4018/978-1-5225-2041-2.ch005

Tiwari, S., Vaish, B., & Singh, P. (2018). Population and Global Food Security: Issues Related to Climate Change. In I. Management Association (Ed.), Climate Change and Environmental Concerns: Breakthroughs in Research and Practice (pp. 41-64). Hershey, PA: IGI Global. doi:10.4018/978-1-5225-5487-5.ch003

Toujani, A., & Achour, H. (2018). A Data Mining Framework for Forest Fire Mapping. In I. Management Association (Ed.), Information Retrieval and Management: Concepts, Methodologies, Tools, and Applications (pp. 771-794). Hershey, PA: IGI Global. doi:10.4018/978-1-5225-5191-1.ch033

Tripathy, B., & K., S. B. (2017). Rough Fuzzy Set Theory and Neighbourhood Approximation Based Modelling for Spatial Epidemiology. In I. Management Association (Ed.), *Public Health and Welfare: Concepts, Methodologies, Tools, and Applications* (pp. 1257-1268). Hershey, PA: IGI Global. doi:10.4018/978-1-5225-1674-3.ch058

Trukhachev, A. (2017). New Approaches to Regional Branding through Green Production and Utilization of Existing Natural Advantages. In I. Management Association (Ed.), Advertising and Branding: Concepts, Methodologies, Tools, and Applications (pp. 1758-1778). Hershey, PA: IGI Global. doi:10.4018/978-1-5225-1793-1.ch081

Tsobanoglou, G. O., & Vlachopoulou, E. I. (2017). Social-Ecological Systems in Local Fisheries Communities. In G. Korres, E. Kourliouros, & M. Michailidis (Eds.), *Handbook of Research on Policies and Practices for Sustainable Economic Growth and Regional Development* (pp. 306–316). Hershey, PA: IGI Global. doi:10.4018/978-1-5225-2458-8.ch026

Tsou, M. (2010). Geographic Information Retrieval and Text Mining on Chinese Tourism Web Pages. *International Journal of Information Technology and Web Engineering*, *5*(1), 56–75. doi:10.4018/jitwe.2010010104

Tuydes-Yaman, H., & Karatas, P. (2018). Evaluation of Walkability and Pedestrian Level of Service. In I. Management Association (Ed.), Intelligent Transportation and Planning: Breakthroughs in Research and Practice (pp. 264-291). Hershey, PA: IGI Global. doi:10.4018/978-1-5225-5210-9.ch012

Uddin, S., Chakravorty, S., Ray, A., & Sherpa, K. S. (2018). Optimal Location of Sub-Station Using Q-GIS and Multi-Criteria Decision Making Approach. *International Journal of Decision Support System Technology*, *10*(2), 65–79. doi:10.4018/IJDSST.2018040104

Uddin, S., Chakravorty, S., Sherpa, K. S., & Ray, A. (2018). Power Distribution System Planning Using Q-GIS. *International Journal of Energy Optimization and Engineering*, *7*(2), 61–75. doi:10.4018/IJEOE.2018040103

Uzun, F. V. (2018). Natural Resources Management. In A. Eneanya (Ed.), *Handbook of Research on Environmental Policies for Emergency Management and Public Safety* (pp. 1–21). Hershey, PA: IGI Global. doi:10.4018/978-1-5225-3194-4.ch001

V, M., Agrawal, R., Sharma, V., & T.N., K. (2018). Supply Chain Social Sustainability and Manufacturing. In I. Management Association (Ed.), *Technology Adoption and Social Issues: Concepts, Methodologies, Tools, and Applications* (pp. 226-252). Hershey, PA: IGI Global. doi:10.4018/978-1-5225-5201-7.ch011

van der Vliet-Bakker, J. M. (2017). Environmentally Forced Migration and Human Rights. In C. Akrivopoulou (Ed.), *Defending Human Rights and Democracy in the Era of Globalization* (pp. 146–180). Hershey, PA: IGI Global. doi:10.4018/978-1-5225-0723-9.ch007

Vaskov, A. G., Lin, Z. Y., Tyagunov, M. G., Shestopalova, T. A., & Deryugina, G. V. (2018). Design of Renewable Sources GIS for ASEAN Countries. In V. Kharchenko & P. Vasant (Eds.), *Handbook of Research on Renewable Energy and Electric Resources for Sustainable Rural Development* (pp. 1–25). Hershey, PA: IGI Global. doi:10.4018/978-1-5225-3867-7.ch001

Vázquez, D. G., & Gil, M. T. (2017). Sustainability in Smart Cities: The Case of Vitoria-Gasteiz (Spain) – A Commitment to a New Urban Paradigm. In L. Carvalho (Ed.), *Handbook of Research on Entrepreneurial Development and Innovation Within Smart Cities* (pp. 248–268). Hershey, PA: IGI Global. doi:10.4018/978-1-5225-1978-2.ch012

Wahab, I. N., & Soonthodu, S. (2018). Geographical Information System in Eco-Tourism. In S. Chaudhuri & N. Ray (Eds.), *GIS Applications in the Tourism and Hospitality Industry* (pp. 61–75). Hershey, PA: IGI Global. doi:10.4018/978-1-5225-5088-4.ch003

Weiss-Randall, D. (2018). Cultivating Environmental Justice. In *Utilizing Innovative Technologies to Address the Public Health Impact of Climate Change: Emerging Research and Opportunities* (pp. 110–143). Hershey, PA: IGI Global. doi:10.4018/978-1-5225-3414-3.ch004

Weiss-Randall, D. (2018). Cultivating Resilience. In *Utilizing Innovative Technologies to Address the Public Health Impact of Climate Change: Emerging Research and Opportunities* (pp. 204–235). Hershey, PA: IGI Global. doi:10.4018/978-1-5225-3414-3.ch007

Weiss-Randall, D. (2018). Climate Change Solutions: Where Do We Go From Here? In *Utilizing Innovative Technologies to Address the Public Health Impact of Climate Change: Emerging Research and Opportunities* (pp. 236–268). Hershey, PA: IGI Global. doi:10.4018/978-1-5225-3414-3.ch008

Whyte, K. P., List, M., Stone, J. V., Grooms, D., Gasteyer, S., Thompson, P. B., . . . Bouri, H. (2018). Uberveillance, Standards, and Anticipation: A Case Study on Nanobiosensors in U.S. Cattle. In I. Management Association (Ed.), Biomedical Engineering: Concepts, Methodologies, Tools, and Applications (pp. 577-596). Hershey, PA: IGI Global. doi:10.4018/978-1-5225-3158-6.ch025

Wulff, E. (2017). Data and Operational Oceanography: A Review in Support of Responsible Fisheries and Aquaculture. In P. Diviacco, A. Leadbetter, & H. Glaves (Eds.), *Oceanographic and Marine Cross-Domain Data Management for Sustainable Development* (pp. 303–324). Hershey, PA: IGI Global. doi:10.4018/978-1-5225-0700-0.ch013

Yener, D. (2017). Geographic Information Systems and Its Applications in Marketing Literature. In S. Faiz & K. Mahmoudi (Eds.), *Handbook of Research on Geographic Information Systems Applications and Advancements* (pp. 158–172). Hershey, PA: IGI Global. doi:10.4018/978-1-5225-0937-0.ch006

About the Contributors

Rajesh Kumar Shakya is the editor of this book. Dr. Shakya is the leading expert in public procurement, e-Government Procurement (e-GP), e-GP Readiness Assessment, Green Public Procurement (GPP), Framework Agreements, Public Investment Management, Digital Governance, and Disruptive Technologies. He has advised the governments in more than 40 nations from Asia, Central Asia, East Asia, Africa, Caribbean, and Europe on public procurement reforms, PFM reforms, e-GP strategies and implementation design, Business Process Reengineering (BPR), Customs and Single Window masterplan development, procurement agency establishment and business model development, Standard Bidding Documents (SBD) preparation, e-Governance solutions design, and introduction of disruptive technologies. He is the architect of the e-Government Procurement Implementation Framework (e-GPIF), which is adapted by most of the countries and international development agencies in implementing the e-Government Procurement (e-GP). Dr. Shakya has been advising the World Bank, USAID, Asian Development Bank, Eastern African Community, Inter-American Development Bank, Caribbean Development Bank, and other government and development agencies for a long time. Dr. Rajesh has got three published books on e-Procurement, several peer-reviewed chapters, e-GP and Framework agreement chapters in Public Administration Encyclopedia, papers on international conferences and seminars, and presentations. Dr. Shakya holds the master's degree with honors in Computer Engineering and the Ph.D. with Distinction in Public Administration from the USA. He also holds professional certifications of PMP from Project Management Institute. He is a member Scientific Committee of International Public Procurement Conference, Working Committee member of European Conference on e-Public Procurement, and Editor and Editorial Panel member of different International journals. Dr. Shakya lives in Maryland, the USA with his wife, a son, and a daughter.

* * *

Daniele Alimonti is a Research Analyst at the Institute of Political Economy and Governance (IPEG) in Barcelona. He provides research support on cutting-edge studies related to public policy and contributes to scientific publications. He consolidated his skills thanks to experiences as a researcher at University Pompeu Fabra and Barcelona Graduate School of Economics (BGSE). His education includes a Master in Economics of Public Policy at BGSE completed after a university career in Economics.

Andrea Appolloni is a Ph.D. graduate and Research Fellow in Management at Tor Vergata University of Rome, Italy. He is responsible for several research projects in Management Consulting, CSR and supply chain related fields such as global sourcing, sustainable supply chains and collaboration. He regularly presents at international conferences and teaches undergraduate, postgraduate and executive level in a variety of subjects including Procurement, Public Procurement, Supply Chain Management, Operations Strategy, and Sustainability. He was the visiting research at Cranfield University, UK to the Logistics e-SCM department and at Tianjin University in China to the Supply Chain and Innovation department. He has worked with a number of companies in his research and professional activities. These include: CTS Voyages in Paris, France, for Hertz rent a car in Dublin, Ireland, Accenture Consulting Company in Italy, Turkey and USA in the field of Sourcing and Procurement Management. He received doctorate degree in Management from the University of Tor Vergata, as well as an MSc in Real Estate Management at the same University and another MSc in International Business Management (focused on China) from Macerata University, Italy.

Maria Antonietta Coppola holds a Bachelor Degree and a Master's Degree in Political Science and International Relations from Sapienza University of Rome. In addition, she holds a second level Master Degree in Public Procurement Management from the University of Rome Tor Vergata.

Ayşenur Erdil is a research assistant at Marmara and Yalova University. She graduated with a BS from Industrial Engineering department, Istanbul University in 2007 and also a double engineering program from Computer Engineering department, Istanbul University in 2008. Then she graduated a Ms from System Engineering program from Yıldız Technical University in 2010. She is a PhD student at Marmara University in Industrial Engineering program at present time. Her current research interests include sustainability, environmental issues, supply chain management systems, production, enterprise resource planning, strategic management systems in business, software selection and implementation in firms, software usability, Computers & industrial engineering, efficiency analysis and quality management systems.

Mehtap Erdil is a graduate with a BS program from Forest Industrial Engineering Department, Istanbul University in 2012. She is an MSc student at Marmara University in Environmental Engineering program at present time. Her current research interests Carbon, water, and energy footprints, environment issues, and, and quality management systems.

Gloria Fiorani graduated with honors in "Economics of Public Administration and International Institutions" in 2004, obtained a Master's Degree in "Innovation and Management in Public Administration" in 2005 and reached the Ph.D. level in "Public Management and Governance" in 2009 (at the University of Rome "Tor Vergata"). Then she was fellowship researcher (SECS P/07), at the University of Rome "Tor Vergata" for 7 consecutive years (from 2009 to 2016). She is currently Adjunct Professor of "CSR and Social Reporting", "Management of Public Administrations", "Social Innovation and Common Goods" and she is Director and Coordinator of Master MARIS – Executive Master in Accountability, Innovation and Sustainability at the University of Rome "Tor Vergata" - Italy, Department of the Management and Law. Her research interests mainly lie in the fields of Sustainable Development, Third Mission in University, Territorial Sustainable Development, procurement, GPP, sustainable supply chain, social reporting.

Ganga G. is an Assistant Professor in Microbiology in Sree Ayyappa College, Eramallikkara, Chengannur, Kerala, India. He has published couple of peer-reviewed papers.

Leslie Harper has over 20 years of experience as an international development professional working on Latin America and the Caribbean. She joined the Inter-American Development Bank (IDB) in 1999, and currently serves as Coordinator of the topic of public procurement modernization and reform. She oversees programs in 26 IDB member countries including portfolio and knowledge management. Ms. Harper has carried out research on public procurement, governance and decentralization, and led knowledge initiatives such as DataGov. Prior to joining the IDB, she worked in a Chilean municipality on a local government strengthening project as part of the Peace Corps program. She has also represented the IDB as a speaker in various international forums in Asia, Europe and Latin America. Ms. Harper has an undergraduate degree from the Elliott School of International Relations at the George Washington University and a Master's Degree in Latin American Studies from the University of London.

Pem Lama has eight years of experience in research and project management in the social and environmental sectors Bhutan. She is currently a consultant at the Asian Development Bank for the Better Business Council Secretariat. Pem held various positions over the 3.5-years working in the Green Public Procurement in Bhutan (GPP Bhutan) project, where she directed research and communication, coordinated project implementation and facilitated training and workshops. She was previously at the Secretariat of the New Development Paradigm, where she was part of the team that produced Bhutan's contribution to post-2015 UN development agenda based on the principles of Gross National Happiness. She started her career at the Bhutan Foundation, the only American non-profit organization that wholly focuses on supporting projects in Bhutan. Pem holds an MBA (Sustainability) from Aarhus University in Denmark and a BA in Economics and International Relations from Tufts University, MA, USA.

Irene Litardi, Ph. D in Management with specialization in Public Management and Governance. She is currently Post-doc fellow in Sustainability and Accountability and teaching assistant in Corporate Social Responsibility at the University of Rome "Tor Vergata", Faculty of Economics, Department of Management and Law, Government of Civil Society Research Group. She is also Program Manager of the Master MARIS – Executive Master in Accountability, Innovation and Sustainability. In 2016, Irene was Visiting Researcher in urban sustainability at the Copenhagen Business School. She graduated in Economics and Management with honors. Since 2015, she has been a member of the permanent study group in "Web Reporting" of the "Social Reporting Group" network and the Secretary of the Department of the "Sustainable Development Solution Network" of the UN. She is interested in Corporate Social Responsibility, Reporting, Territorial Sustainable Development, Urban Regeneration and Resilience, Social Engagement, Green Publi Procurement and sustainable supply chain. She is ISIPM certified project manager (ISIPM-base).

Gustavo Piga, Ph. D. in Economics at Columbia University, is Full Professor of Economics at the University of Rome Tor Vergata, where he chairs the Master in Procurement Management. He has chaired the Italian Procurement Agency for Goods and Services, Consip Ltd., between 2002 and 2005. His fields of expertise are public procurement, macroeconomics and public debt management. He has co-edited several books, among them the Handbook of Procurement, Cambridge University Press, with Nicola Dimitri and Giancarlo Spagnolo and Revisiting Keynes: Economic Possibilities for our Grandchildren, MIT Press, with Lorenzo Pecchi. He is co-editor in chief of the European Journal of Public Procurement Markets.

Daniel Sánchez is a Senior Public Procurement Consultant with over 15 years of experience in international development. He currently consults for the Inter-American Development Bank (IDB), developing original research in the area of public procurement. Mr. Sanchez has worked with several countries in Latin-America and the Caribbean, to accelerate reform and modernization of their government procurement systems. He has also participated as a speaker in various international procurement forums in USA, Europe and Latin-America. Mr. Sanchez a Bolivian national, holds a MBA from the joint program sponsored by the Harvard Institute for International Development and the Bolivian Catholic University. He also holds an MA in Public Communication from American University and has carried out post graduate studies on public financial management at the Harvard Kennedy School of Government.

Index

L

M

O

P

R

S

T

V

W

CPSIA information can be obtained
at www.ICGtesting.com
Printed in the USA
BVHW060305050619
550182BV00008B/69/P

9 781522 570837